T0072425

ALSO BY AMY SUTHERLAND

Kicked, Bitten, and Scratched

Cookoff

WHAT SHAMU TAUGHT ME ABOUT

LIFE,

LOVE,

AND

MARRIAGE

WHAT SHAMU TAUGHT ME ABOUT LIFE, LOVE, AND MARRIAGE

LESSONS FOR PEOPLE FROM ANIMALS AND THEIR TRAINERS

AMY SUTHERLAND

RANDOM HOUSE TRADE PAPERBACKS / NEW YORK

2009 Random House Trade Paperback Edition

Copyright © 2008 by Amy Sutherland

All rights reserved.

Published in the United States by Random House Trade Paperbacks, an imprint of
The Random House Publishing Group, a division of Random House, Inc., New York.

RANDOM HOUSE TRADE PAPERBACKS and colophon are trademarks of
Random House, Inc.

Originally published in hardcover in the United States by Random House, an imprint
of The Random House Publishing Group, a division of Random House, Inc., in 2008.

Portions of this work were originally published, in different form,
in *The New York Times*.

LIBRARY OF CONGRESS CATALOGING-IN-PUBLICATION DATA
Sutherland, Amy.
What Shamu taught me about life, love, and marriage : lessons for people from
animals and their trainers / Amy Sutherland.
p.   cm.
ISBN 978-0-8129-7808-7
1. Typology (Psychology). 2. Animals—Psychological aspects.
3. Interpersonal relations. 4. Human-animal relationships. 5. Animal training.
I. Title.
BF698.3.S755 2008
158.2—dc22   2007037869

www.atrandom.com

Book design by Casey Hampton

146122990

For the animals, especially
the human animal I
married, Scott

CONTENTS

INTRODUCTION

Sitting at my desk in Maine this summer morning, I hear my elderly neighbor clear his throat with undue enthusiasm as he trundles down his driveway. A foghorn moans. A car door slams. I hardly notice these ordinary, everyday sounds as my eyes are trained on the extraordinary, a delicate underwater dance a continent away. A black figure, a swimmer in a wet suit, I can't tell if it's a man or a woman, splashes in the middle of a deep pool. Behind the swimmer, a killer whale glides through the tank. His dorsal fin stands tall like a sail. His sharp black-and-white markings reflect in the water's surface above.

Thanks to a webcam, I can see the shimmering show tank at SeaWorld San Diego. That is, I can see the seven-million-gallon pool from the surface down, an altogether different view from that of anyone actually at Shamu Stadium just now—what I would call a crab's-eye view. This is a very blue world. The chilly water is the color of lapis, but the shade brightens and then deepens as the sun arcs overhead. The rocks on the tank floor are the indigo of a starless night sky. Aquamarine light crackles across the tank's white bottom as the water tosses to and fro.

Even the killer whale casts a blue shadow as he circles the tank. Today there is just one whale in the water, but I've seen two, even three. I've seen babies. I've watched them scratch their backs along the rocks and gush balloon-sized bubbles. Some like to lap the pool upside down, their alabaster bellies turned skyward. One has white markings in the shape of a ginkgo leaf. When the whales swim close to the camera, and their stomachs suddenly fill the frame, I can't help but whoop. My dogs will lift their heads and turn their bright, curious eyes to me. Downstairs at his desk, my husband will call "What?"

Nothing, though, compares to when a trainer is in the tank. Their little legs kicking and arms waving about, the trainers look like water bugs compared to the sleek eight-thousand-pound whales. I can't tear my eyes away from the sight of these two wildly different creatures working in tandem. And so I remain glued to the screen as the whale turns its bulk toward the swimmer and nudges its broad black rostrum under his feet.

———

I admit it. I'm procrastinating. I should be writing, actually, this very book. But this cross-country cyberview of a multiton animal in synch with a relatively itty-bitty human is a very apropos way to dither. It reminds me how much I have changed.

I'm an altogether different person than I was three years ago. My friends and family may not have noticed, but I am almost unrecognizable to myself at times. My outlook is more optimistic. I'm less judgmental. I have vastly more patience and self-control. I'm a better observer. I get along better with people, especially my husband. I have a peace of mind that comes from the world making so much more sense to me.

What brought about this change? Counseling? Nope. Happy pills? Nope. Yoga? Nope. A religious awakening? Wrong again. Acupuncture? Definitely not.

I discovered a school for exotic animal trainers, and wrote a book about it. That's what.

Funny thing is, I wasn't looking to change, but change has always had a way of finding me. I learned early in my journalism career that whatever I wrote about, whether it was blueberry farming or avant-garde jazz, eventually got under my skin to some degree. When I wrote a book about the world of competitive cooking in America, before long I was dreaming up recipes and submitting them. When I worked on a series about domestic violence, I began to have nightmares. If I was this easily influenced, I decided, I'd have to stay clear of darker topics. Complicated was okay, deeply troubling was not.

For my second book I followed students at Moorpark College's Exotic Animal Training and Management Program, the Harvard University for animal trainers, for one year. I walked away with not only a California tan, a new respect for scavengers, and more than enough material for a book, but something I never expected—a whole new approach to life.

While I worked on that book, I droned on to friends and family, anyone who would hold still for thirty seconds, of the great wisdom I had found at the feet of animal trainers. My friends, who had listened to me rattle on about how to cube meat for competitive chili while I wrote *Cookoff*, nodded good-humoredly and, I suspect, thought to themselves, "There she goes again." Sometimes they'd interrupt me to ask "When is your book due?" in hopes the deadline was not far off and that I'd soon be on to another topic.

My husband, who, like me, loves animals and knows his fair share about training, was far more receptive, and didn't even make fun of me when I began tossing about terms like "incompatible behavior" or "instinctive drift." But not even he really understood at first what I was up to—that I had begun to use the animal training techniques, only I was using them on my fellow species, and on no one more than my handsome husband.

Eventually, I wrote a column for *The New York Times* about how I had improved my marriage by thinking like an animal trainer. To my surprise, the whole world sat up and took notice. After being ignored by my friends, I was suddenly besieged with interview requests from around the globe— Brazil, Ireland, Spain, Canada. Four reporters called me from Australia alone. My in-box filled up with congratulatory e-mails. I landed on the *Today* show. Hollywood called. My column shot to the top of the list of most e-mailed stories at the *Times*, where it remained for days, then weeks, and eventually became the most e-mailed story of 2006. When the dust settled, I had a movie deal and a contract to make my *Times* column into a book.

I never expected to write this book, or anything like it. But then I never expected that animal training would transform me. I'm not a counselor or a minister or a trainer or an expert of any kind. I'm a journalist. What I have to offer is my own personal story as a kind of Alice who stumbled into a Wonderland where cheetahs walk on leashes, hyenas pirouette on command, and baboons skateboard, and left with a new outlook on marriage, men, humans, life. My experience may give you some food for thought, a laugh, a light dose of philosophy, a way to solve some small problems that aren't worth a visit to the shrink but still nag. Or it may change you from head to toe.

The world is full of surprises. The proof is before me.

With the whale's snout under his feet, the trainer pulls his arms flush to his sides as Shamu pushes him through the water. It is a magical sight, even more so as the twosome zooms past the webcam. The trainer's head cuts through the water and comes into view first. Now I can see the trainer is a woman. A blond ponytail streams down her back. Her face turned forward, she looks like a ship's figurehead. A trail of small bubbles escapes from the corner of her mouth. Her straight, horizontal body follows. Then I see her feet, which are still neatly balanced on the rostrum of the ocean's top predator. Shamu's sleek two-ton bulk fills the camera's lens and then disappears. Though I can't see the pair, I know what they are doing, bursting from the tank like water gods at whom a stadium of onlookers will scream and clap and marvel, like me at my desk, at what seems impossible but isn't.

WHAT SHAMU TAUGHT ME ABOUT
LIFE,
LOVE,
AND
MARRIAGE

ONE

PEOPLE ARE ANIMALS TOO

s I wash dishes at the kitchen sink, my husband paces behind me, irritated. "Have you seen my keys?" he snarls, then huffs out a loud sigh and stomps from the room with our dog, Dixie, hot on his heels, anxious over her favorite human's upset.

In the past I would have been right behind Dixie. I would have turned off the faucet and joined in the hunt while trying to soothe Scott with cheerful bromides like "Don't worry, they'll turn up!" Sometimes I'd offer wifely pointers on how not to lose his keys to begin with. Or, if I was cranky, snap "Calm down." It didn't matter what I did, Scott typically only grew angrier, and a simple case of missing keys would soon become a full-blown angst-ridden drama starring both of us and Dixie, our poor nervous Australian shepherd. Penny Jane, our composed border collie mix, was the only one smart enough to stay out of the show.

Now, I focus on the wet plate in my hands. I don't turn around. I don't say a word. I'm using a technique I learned from a dolphin trainer.

I love my husband. With his fair skin and thick chestnut hair, he's handsome in an angular Nordic way. He's well read and adventurous, and does a hysterical rendition of a northern Vermont accent that still cracks me up after fourteen years of marriage. We like many of the same things: dogs, jazz, medium-rare hamburgers, good bourbon, long walks, the color orange. But he can also get on my nerves. He hovers around me in the kitchen when I'm trying to concentrate on the simmering pans, asking me if I read this or that piece in *The New Yorker*. He finishes off boxes of cookies, especially the dense caramel bars his mother sends from Minnesota, then says "I thought you were done with them." He leaves wadded tissues in the car. He drives through red lights, calling them "long yellows." He suffers from serious bouts of spousal deafness, yet never fails to hear me when I mutter to myself on the other side of the house. "What did you say?" he'll shout. "Nothing," I'll yell back. "What?" he'll call again.

These minor annoyances are not the stuff of separation and divorce, but in sum they dulled my love for Scott. Sometimes when I looked at him I would see not the lean Minnesotan I adored but a dirty-Kleenex-dropping, hard-of-hearing, prickly cookie monster. At those moments, he was less my beloved husband and more a man-sized fly pestering me, darting up my nose, landing in the sauce on the stove, buzzing through my life.

So, like many wives before me, I ignored a library of advice books and set about improving him. By nagging, of course, which usually had the opposite effect from the one I longed for—his size 11 shoes continued to pile up by the front door, he went longer between haircuts, he continued to return empty milk cartons to the fridge. I tried cheerful advice like "You are so handsome, but no one can see it behind your five o'clock shadow." That usually resulted in another couple of razorless days. I made diplomatic overtures like "What if we each promise not to leave smelly clothes lying around?" "Okay," my husband would agree good-naturedly, and then walk right past his reeking bike garb on the bedroom floor.

I, a modern woman, tried being direct, asking him in a voice as neutral as a robot's, "Would you please not drive so fast?" Even this approach would backfire as in my simple question my husband might hear an accusation or an order and then push the accelerator a hair more. When all else failed, I yelled, and then we fought.

We went to a counselor to smooth the edges off our marriage. The counselor, a petite, sharp-boned woman who took notes on a legal pad, didn't understand what we were doing there and complimented us repeatedly on how well we communicated. I threw in the towel. I guessed she was right—our union was better than most—and resigned myself to the occasional sarcastic remark and mounting resentment.

Then something magical happened. I discovered animal training.

I stumbled into the world of animal training nearly ten years ago when we brought home Dixie, an eight-week-old herding dog, ten pounds of furry red energy. It was as if we had lit a bottle rocket in our house, the way she ricocheted from room to room, a toy or two hanging out of her mouth. I gave up meditating in the morning to begin my daily pursuit of wearing her out. It was a sunup-to-way-past-sundown job. Before I even got dressed or made coffee, I would sit cross-legged on the floor, hold a faux sheepskin rug before me, and call "Get it." Dixie would catapult herself into the rug and rip it from my hands, her amber eyes afire, and then we'd each tug with all our might. We played that game so much, the rug was eventually reduced to a slobber-encrusted handful of fabric.

I learned to throw a ball properly for the first time in my life, and then a Frisbee. I tossed balls and Frisbees and walked so much I went down a size in pants. Dixie was either tugging, wrestling, or running, or she was fast asleep under a table where we couldn't pet her. Should we get down on our hands and knees and reach under to pat her, Dixie would look miffed, like an Olympic athlete roused from a power nap, then pull herself to her paws and move just out of reach. Cuddling, from Dixie's point of view, was for wuss dogs.

Though I think we were a bit of a disappointment to Dixie, the way commoners can be to royalty, we were just smart enough to know that a herding breed needed a job. So we went looking for an agility class, where you learn to run your dog through a whimsical obstacle course of tunnels, jumps, and teeter-totters. At that time, we found only one trainer around Portland, Maine, who taught this crazy skill. Before we could tackle the course, though, we were required to take a puppy training class.

If this trainer had used traditional techniques, the leash-popping and pushing your dog this way and that, I think the story would have ended there. For me, there is little magic nor imagination in that old-school approach. But it was my good luck that the trainer used progressive, positive techniques, techniques based on an altogether different philosophy. Rather than learning to boss our pups around and make them into obedient dogs, we learned to communicate and cooperate with them. She didn't teach us just how to get our dogs to sit, but rather how to think about our canine companions.

Amid the joyful chaos of puppy class—the barking, the tangled leashes, the marital squabbles—I found an intellectual and personal challenge I hadn't expected. I found a new me, a me with much more patience and self-control. I learned to be precise and observant. I learned to teach Dixie what I wanted rather than what I didn't want. I learned not to take anything she did personally, not even when she ripped my shorts in a fit of overexcitement. All this from a six-week puppy training class.

I also began communicating with another species, and you can never underestimate the thrill of that. I signed us up for another class, and another class, and another class. I was hooked. So hooked that when I landed on the Paris set of *102 Dalmatians* for a magazine assignment, I spent every spare moment hanging out with the animal trainers, chatting about such things as how they taught a parrot to ride atop a bull-mastiff and how they got the dog not to shake whenever the bird's wings brushed its back. The trainers, to my surprise, had all earned actual degrees in exotic animal training. They had studied at a community college outside Los Angeles. It was the go-to school, they told me, not to mention the only program of its kind. Back home, I taught Dixie to bring in the Sunday *New York Times*, scribbled down the name of the school, and threw the scrap of paper into my idea folder.

———

In 2003 I began work on a book about this school. For a year, I commuted between Maine and California, where I followed students at Moorpark College's Exotic Animal Training and Management Program. There I spent my days watching students do the seemingly impossible: teach a caracal to offer its paws for a nail clipping, a camel to shoot hoops, a baboon to get into a crate and close the door behind her. Each day at the teaching zoo was packed with countless lessons, from how to pick up a boa constrictor to how to speak to a wolf. As I observed the students, I essentially became a student too. I learned not to look the primates in the eye, to stride with confident ease while on a cougar walk, and never to stand close to any enclosure, especially not the big carnivores'. I learned that when Zulu the mandrill bobbed his head at me, he was saying "Back off." That when Rosie the baboon smacked her lips together, she was saying "Hello, friend." That when Julietta the emu made a thumping noise in her chest, she was worried.

I learned the language of animal trainers, what they meant when they talked about A-to-Bs (teaching an animal to move from one spot to another) or targeting (having an animal press its snout to something). If somebody told me that they had just been grooming with the squirrel monkey, I knew that they had sat close to the cage, held up their arms, and let the monkey run his black fingers over their skin. I learned what a *positive count* is (making sure the animal is there in its enclosure) and that B.E. stood for "behavioral enrichment," basically anything that made the animal's life more stimulating, whether it be a toy or a walk on a leash. Training, it turns out, is one of the things that make an animal's life interesting. So you could even teach an animal an A-to-B for B.E.

I soaked up their sayings, such as "Go back to kindergarten," shorthand for when an animal has trouble learning a behavior and the trainer needs to back up a few steps in the training. "Train every animal as if it's a killer whale" meant to work with every animal as if you could neither forcibly move it nor dominate it. "It's never the animal's fault" is pretty much what it says: If an animal flounders in training, it's the trainer's fault. One of my personal favorites was "Everything with a mouth bites." I wrote down that line in all caps, for my research but for myself too. Why? I wasn't sure exactly. It had a philosophical ring. It was also silly, but made such good, plain sense, a funny reminder of what a great leveler Mother Nature can be. A cute, fuzzy animal will bite you just as easily as a mean-looking one. By the same token, the animal doesn't care whether you're as angelic as Mother Teresa or as loathsome as Caligula. Shiny auras, the best intentions, and sainthood don't mean much, if anything, in the animal kingdom.

So much of what I was learning at the school had meaning beyond the front gate. This place, where the great divide between animals and humans closes, captured my imagination in a way nothing else had. Every visit drove home how complicated, weird, and fantastic the natural world is. I felt my mind crack wide open trying to take it all in.

I trailed the students to class and then out to the teaching zoo grounds where they practiced on a badger or a lion or the mysterious binturong, a rain forest animal that resembles a raccoon on steroids. I watched as one student trained an olive baboon to let her rub lotion on her hands, as another taught a capuchin monkey to unravel its long leash when it became tangled during walks, and yet another instruct the Bengal tiger to get in her kiddie pool on command. I tagged along on field trips, during which I listened, rapt, as professional trainers explained how they taught dolphins to flip or ibis to fly to them. At a private compound in Southern California, I took notes in the fading light of day as six elephants, on command, lined up in a row, urinated, turned to their left in unison, hooked trunks to tails, and, single file, swaggered into the barn for the night. In Cincinnati, I saw a leashed cheetah sit calmly on a desk next to a trainer as she lectured to a bewitched audience. At a conference in Baltimore, I listened as trainers described how they had taught spotted eagle rays to swim to a feeder.

I can't say the instant it happened, but eventually it dawned on me that if trainers could work such wonders with spotted eagle rays and baboons and dolphins, might not their methods apply to another species—humans? It was not much of a leap for me. By just watching, thinking, and reading about animal behavior, I had discovered a good deal about the behavior of my own species. In a kind of reverse anthropomorphism, I couldn't help seeing parallels, especially with the primates, but with all the animals, even the turkey vulture, which, like us, sunbathes. True, people are more complicated than animals, but maybe not as much as we assume. As the relatively nascent field of animal behavior continues to grow, more and more research shows that animals are anything but mindless organisms driven solely by instinct. Traits that were considered unique to humans, such as tool use and collaboration, have been found among other primates and now birds and fish. Turns out groupers and moray eels hunt together, and crows are quite handy with a bit of wire.

Complicated or not, we *Homo sapiens*, the highest of the primates, the tippy-top of the food chain, a frighteningly successful species, are, in the end, members of the animal kingdom, like it or not. Animal trainers showed me that there are universal rules of behavior that cut across all species. Why should we be any different?

I began to take home what I learned at the teaching zoo. If my husband did something that annoyed me, I thought, "How would an exotic animal trainer respond?" If I got into a squabble with a relative, I did the same. If the clerk at the post office gave me a hard time, likewise. That may sound ridiculous. I admit it. In fact, at the start I thought of it as a kind of goofy experiment, but the early results were so convincing that I kept at it.

Did I teach my husband and friends to sit and stay? No, of course not. What would be the point of that? Okay, it would be a funny party trick, especially if I trained them to crow and scratch like roosters on command. My purpose, however, was not to bend people to my will, but to better navigate the human interactions and relationships that fill my days. Funny thing is, I ultimately learned some pretty obvious lessons, such as to have more patience with my husband, with everyone for that matter, lessons I could have gotten out of a self-help book or in half a session with a therapist. But had I heard these words of wisdom from a counselor or read them in some overearnest manual stamped with smiley faces, I would have thought, "Duh," and gone off and lost my patience with someone. Even if I had cracked a dolphin-training manual, I still wouldn't have been inspired to change. But learning, actually seeing, these obvious lessons via sea lions, fennec foxes, Harris's hawks, and squirrel monkeys captured my imagination and made self-improvement, for once, engrossing, even fun. Instead of thinking, "Don't call him a space shot," I'd ask myself, "What would a dolphin trainer do?"

That is what I'm doing right now as my agitated husband looks for his keys.

———

The answer is "nothing." Dolphin trainers, in fact all progressive trainers, reward the behavior they want and ignore the behavior they don't. So I'm ignoring behavior I don't want— Scott's rising temper. I don't even call out places to look. Rather, I, lips sealed, keep at the task at hand, rinsing a plate. At the sink, I hear my husband bang a closet door shut, rustle through papers on a chest in the front hall, and thump upstairs. I pop the plate into the dishwasher and rinse another. Then, sure enough, all goes quiet.

A moment later, Scott strides into the kitchen, keys in hand, and says calmly, "Found them."

Without turning, I call out, "Great, see you later."

Off he goes with our much-calmed pup. The drama averted, I feel like I should toss him a mackerel, maybe toss myself one too. It's not easy thinking like an exotic animal trainer.

TWO

ANY INTERACTION IS TRAINING

When I told the world that I had used animal training techniques to improve my marriage, I got a few irate e-mails from men who accused me of manipulating my husband, not to mention demeaning him by comparing him to an animal. To me, the latter complaint is moot, because humans *are* animals, even husbands, even husbands with the IQs of rocket scientists, even rocket scientists. Besides, I never met a man who minded being compared to a lion or tiger, even a bear. And wouldn't we all like to be a little more like dolphins, who, with their brains, good looks, and athleticism, are the Kennedys of the animal kingdom? I dated guys who couldn't hold a candle to a dolphin.

As to the first point, I turn again to the world of animal trainers, though I realize this is what got me in hot water to begin with. Specifically, I turn to one of their sayings: Any interaction is training. Translation: Every time you have any kind of contact with an animal—when you leave food on the floor, talk to it, pass its enclosure—you are teaching it something whether you mean to or not. Even just looking at an animal can be an inadvertent lesson. Animals never stop learning from their environment, and if you are a part of that environment, they will pick up a thing or two from you and your behavior. If Mother Nature hadn't made them so, they would have long since slid into the evolutionary abyss. That animals are designed to learn is the reason trainers can train them.

Animals are such quick studies that trainers have to be careful. A thoughtless yelp or poorly timed reward, and an animal is suddenly doing something no one planned on. In his definitive book *Animal Training*, Ken Ramirez, who oversees the training at Chicago's Shedd Aquarium, recounts how some beluga whales learned to spit water at the trainers. Whales are natural spitters. In the wild, the whales spit on the ocean floor to uncover food, such as sandworms, snails, or flounder. They've also been spotted spitting above the ocean's surface, at a floating log, say, or into the air like a fountain. So it wasn't a huge leap for a beluga at the Shedd to shoot a mouthful of chilly tank water at a passing wet suit, whereupon the surprised trainer probably squealed or jumped or both. Another beluga in the tank gave it a try, resulting in more squeals and jumps. A third joined in. The trainers had unintentionally trained the belugas to be whale-sized squirt guns.

This is why students in their first year at the training school are forbidden to have any interaction with the animals—

not even talking or eye contact—lest they unwittingly teach the animals unwanted behaviors. Still, the newbies innocently taught Rosie the olive baboon a thing or two. The baboon learned to sit calmly and smack her lips in greeting at the new students, which would lure them close to her cage. Then, just as they cracked a smile, Rosie would shriek in their fresh faces. The more the students would jump and squeal, the surer Rosie was to try this trick again. The cockatoo Clyde came up with a similar ploy. The white bird, sitting nicely on his perch, would look as friendly as a welcoming committee when new students stepped into his cage to clean it. "Hi, Clyde," he'd squawk, and suddenly fly at their heads, claws out, wings flapping. The students would duck and scream, which Clyde obviously enjoyed. The camels, Sirocco and Kaleb, adopted a silent and more subtle ruse. The camels figured out that if they leaned out of their corrals at the same time, they could, with their long, thick necks, trap new students in the space between their enclosures. Thus captured, the newbies, who were forbidden to talk to or touch the camels, would have to wait for a veteran student to happen along and free them. In the meantime, Sirocco and Kaleb had a nervous, embarrassed hostage, which apparently has its appeal to camels.

Even before I officially began training our second dog, Penny Jane the border collie mix, I taught her some lessons without meaning to. At some point in her early days with us, she barked on the back porch. Not wanting her sharp woofs to bother our neighbors, especially the couple with the baby just behind us, I rushed to let her in. Penny Jane made a mental note. In short order, whenever she wanted in, she barked.

Penny Jane even developed her "Let me in" bark, a single, short, emphatic, some might even say snotty, *arf,* which I can hear in the farthest corners of the house, usually when I'm on the phone, in the bathroom, or otherwise right in the middle of something.

I also taught her, by accident, that I am a menace with a car door. One night, we loaded Dixie and Penny Jane in the backseat for a short ride to the local video store. I, feeling warm and fuzzy from a nice big Scotty-made martini, did not notice that Penny Jane had unfurled her usually curled tail. I slammed the door closed, catching the white tip of her lovely tail. Penny Jane yelped. I opened the door as fast as I could. Her tail, miraculously, was fine. Her nerves were not. She learned to keep her distance when my hand was on the car door handle. To this day, whenever she gets in the car, Penny Jane bolts for the far side of the seat and tucks her tail safely under her behind. I learned that after one martini I can't keep track of whose tail is where.

So you don't have to purposely train an animal to train it. Even at old-fashioned zoos that eschew training, believing that the animals should be kept as "wild" as possible, the animals still learn a long list of behaviors regardless. If keepers feed the animals at the same hour every morning in the same spot, the animals learn the breakfast bar opens at 8 A.M. by the moat. Zoo animals quickly realize that when a vet shows up they'd best hide. Or that if they bang long enough and loud enough on the enclosure door, the keepers will come running, maybe even offer bananas to stop all the noise.

Zookeepers unschooled in training can even unwittingly encourage polar bears to pace by trying to stop them. The bears are given to pacing in captivity. Nobody knows exactly why. The repetitive padding may be soothing or stimulating. It may be for exercise. In the wild the bears walk, heel to toe just like us, over miles and miles of ice and tundra in search of seals and other snacks. What zoos do know is that a pacing bear upsets visitors. Zookeepers will toss in a big ball to distract an animal from compulsively parading back and forth in its enclosure. To the bear, though, that may seem like a nice reward for *pacing*. Next time the bear gets a hankering for a toy, he puts one huge white paw before the other, and voilà, a ball appears.

Even zoo visitors can teach the animals. A few years back at my hometown zoo, the Cincinnati Zoo, a twentysomething female gorilla by the name of Muke pitched a chunk of sod across the moat at the gathered onlookers. Chaos, amused and otherwise, erupted. She threw another chunk. Screaming, running, laughing ensued. In short order, visiting the gorillas became like a one-sided game of dodgeball. Some people thought it was funny. Some people experienced post-traumatic stress disorder as the scene brought back bad gym class memories. Muke, whose aim improved with practice, eventually nailed a little girl on the head with a clump of turf, giving her a knot and a story to tell at cocktail parties for the rest of her life. The zoo posted keepers by the enclosure to warn of incoming sod balls and to inform visitors that they weren't allowed to throw back. Oh, yeah, we're nothing like animals.

———

Animal trainers argue, why not teach animals purposely? Why leave what they learn to chance? The same question, it occurred to me, is worth asking about humans.

All of us, whether consciously or not, spend a good chunk of our day trying to alter each other's behavior. When you tailgate someone, you hope to make the car ahead speed up or get the hell out of the way. When you help someone, say, by explaining the appropriate length for toenails to your spouse or teasing a friend for obsessively tracking the underwear habits of Britney Spears, you are, to some extent, trying to change them. When you do something just right, say, organize your recycling bin perfectly or keep your lawn just so, you offer a model, consciously or unconsciously, in hopes your neighbors will do things your way. When you sigh with the drama of a movie star while waiting for a table at a restaurant, you are trying to get the host to seat you immediately. We may be direct or indirect, polite or pushy, but we are all aiming for the same bird.

This bent, I think, is most obvious between parents and children, who are typically caught in such a dizzying behavioral cycle you can't tell who is training whom, but it goes on as well between mates, friends, coworkers, relatives, and strangers nonstop. Along the way, we teach plenty of behaviors by accident. How many times has a spouse whined, "But I thought you liked it when I (fill in the blank)." Parents unintentionally teach their kids to resist going to bed, as the more the youngsters dig in their heels, the more mom and dad negotiate, plead, and bribe. People who respond to lunch and dinner invitations with breathless e-mails listing their busy schedules train their friends not to invite them to anything. Many employers encourage the exact opposite of what they want—doing the bare minimum—by not rewarding workers who go the extra mile and by heaping lots of attention, albeit negative, on the slackers. Even institutions are in on the game. The airlines may have saved themselves financially in recent years, but along the way they have taught a nation that loved flying to hate it.

I am no exception. I had been unwittingly training people my whole life, none more than my husband, Scott, and he me. My techniques were nagging, occasional diplomatic overtures, pleading, sarcasm, and a personal favorite, the cold shoulder. His were spousal deafness, occasional snarls, edicts ("No more loud nose-blowing," he proclaimed once), and more spousal deafness. We rarely got the results we wanted. Along the way, I had, by mistake, trained him to take refuge in the bathroom every time I mentioned gardening. He had trained me to honk my nose even louder (it was fun to see him cringe). Though none of our techniques worked all that well, we persisted like old-school circus trainers who knew only one way to work an animal, mostly a negative way.

So if I was already essentially trying to change my husband, not to mention friends, relatives, coworkers and whomever, why not do so consciously, not to mention effectively? Why not think like an animal trainer, like a progressive animal trainer? Why not change behavior on purpose rather than by accident?

This is just how purposeful animal trainers are. They pick a behavior they want to train, say, a sea lion's balancing a ball on its nose, a classic. Next they come up with a step-by-step procedure to that end: Hold the ball out for the sea lion to feel with its strong blond whiskers, hold the ball overhead so the animal stretches its rubbery neck and whiskers straight up, get the animal to hold that position, and finally, set the ball on its moist nose. Trainers write all these steps down. Then they note their progress from one training session to the next. Marine mammal trainers especially can churn out page after page of this stuff, even lovely graphs.

Now, I'm not nearly that organized, so a written training plan was not realistic for me. And my training would always be on the fly as I rushed through the day. Animal trainers have a big advantage that I don't: It is their job. When they train, that is all they are doing. They aren't answering the phone, looking for a yogurt in the fridge, or paying bills while checking to see if a dolphin correctly slapped the water with its pectoral flipper. With all the distractions of everyday life, I would never have the concentration of a pro. Not to mention that my animals could talk back, saying things like "Are you using animal training on me?" But it was still worth a try.

So, plan or no, like a trainer, I needed to set a few specific goals, to think of what I wanted and didn't want, not just keep stumbling along. As it was, I was working like a trainer who just tosses a ball to a sea lion to see what will happen. If I wanted the ball balanced on that whiskered nose, figuratively, I needed to be far more deliberate.

THE T-WORD

Uncomfortable? Though we train for marathons and hire personal trainers by the thousands, lots of people are uneasy with that word—*training*—when it comes to humans. Annie Clayton found out how uneasy the hard way.

In 2005, the BBC aired *Bring Your Husband to Heel*, a reality show that featured Clayton, who is a dog trainer. Clayton showed beleaguered wives how to use the basic principles of progressive dog training to improve their spouses' annoying habits. Clayton taught the wives to reward behavior they liked and to ignore what they didn't. This subtlety was lost on viewers, who complained that the show was demeaning, not to mention "scandalous drivel," as *The Guardian* reported. This in a country known for its great love of our four-legged best friends. In short order, the BBC canceled the series, issued a formal apology saying it was sorry for "any upset" the show might have caused, and threw Clayton to the blogosphere.

The show's ham-handed title didn't help, but Clayton was, to some degree, the victim of the very negative associations some people have with "training." Maybe potty training didn't go so well for them, or they can't afford personal trainers. Regardless, to many folks the word doubles for "manipulation" or "control." It conjures up images of big-cat trainers with a whip in one hand and a chair in the other, of wild animals being tamed, their noble souls bent to man's will.

Though training has come a long way, this out-of-date image is, unfortunately, still very much alive in people's minds. Most people's ideas of training are based on old-school approaches, which have been passed down from the caveman and are, sadly, still with us. This is the style that, until recently, dominated the dog world. With traditional training, the goal is to have an animal do as it is told, to break it, show it who's boss. This mind-set alone is enough to give people a negative idea of training. Who wants to be broken?

Moreover, with this approach, whether an animal enjoys training is typically immaterial. The animal must do as it is told or else. Traditional trainers may use rewards or not, but they always use punishment, from the token (taking away a toy) to the heavy-handed (a crack on the snout). They teach an animal what not to do. If a dog pulls on its leash, pop its choke chain. If it doesn't sit when told, *pop.* Doesn't lie down fast enough, *pop.* An animal is motivated by wanting to avoid something bad, a pinch, crack, or zap. This style exploits what animals don't like. There is something oh so human about that.

Progressive trainers take a fundamentally opposite approach. They think of training as communication. They teach rather than tame. They don't *make* animals do anything, but rather entice them. Their goal is to make animals not obedient, but engaged. They want animals to like—no, *love*—training. That rules out pinches, cracks, and zaps. Purists won't even use the word "no." It's not that these trainers wouldn't pull back on the leash or shout if a dog lunged for the Thanksgiving turkey, but they don't use punishment as a teaching tool. They motivate animals only with rewards. Cue a dog to sit. If his bottom hits the floor, he earns a treat. If he doesn't take a seat, there are no repercussions except not getting kibble. This way, an animal never has anything to lose.

This enlightened philosophy comes from marine mammal trainers and has caught on with more and more trainers of all species, from hornbills to hippos. These are the principles behind clicker training, which has revolutionized dog training. This fresh approach is also what has inspired zoos to reconsider training and aquariums to try it with fish, turtles, and octopuses. It has changed how training is used, to improve the lives of captive animals. Animals in zoos and aquariums are trained in order to give them physical exercise, to provide mental stimulation, to help with their own care, and to solve problems. Say a zoo capuchin is pulling out its own fur from boredom, the monkey equivalent of playing too many video games. In response, the keepers teach—yes, train—the capuchin all kinds of new behaviors: walking on a leash, even a game of peekaboo. Soon the monkey is too busy, not to mention tired what with all the activity, to obsess over its fur. Problem solved, and without one single antidepressant. As Gary Priest, who oversees training at the San Diego Zoo, says, "Where there's a problem, there's a behavioral solution."

Working with rewards, trainers have also been able to teach species and behaviors previously thought to be untrainable. At Gatorland in Florida, the reptiles have been taught to burst out of their murky pond on command. Trainers at Saguaro National Park in Arizona tell free-flighted raptors which way to fly by looking in that direction. At the Baltimore Aquarium, the tamarind monkeys give urine samples on cue.

These are the kind of trainers I spent hours and hours with by various tanks and enclosures. This is why, for me, "training" is a magical word, so magical in fact that I can forget how negative it is for other people. If I do, forgive me.

WALKING THE WALK

Why not just be direct, as some people put it to me, and tell people what you want or don't? First, why speech is automatically considered a more direct means of communication than behavior is a mystery to me. Second, talk isn't all it's cracked up to be.

We humans love language to a fault. We blab on about how speech makes us superior communicators to animals. Really?

Perhaps animals aren't so verbal, but they may have it over us in the clarity department. Granted, animals usually have something simple to get across, such as "I'm the boss of you," "No, you're not," "Let's get it on now," "Bananas this way," or "Leopard!" But they manage all that, and far more, without a single word. True, animals cannot, like us, communicate all that plus discuss Proust, debate stem cell research, and parse our feelings ad infinitum. But keep in mind that if it was to a species' evolutionary advantage to discuss the madeleine reverie in *Remembrance of Things Past*, it would have evolved to do so. Mother Nature would have supplied the needed DNA. As it is, we are the only species to need that skill—so far. And sometimes we are so busy debating nineteenth-century French literature that we don't yell "Leopard!" when we should.

That we can talk gives us an edge over animals in communication, but we so often sabotage ourselves with our own language. Despite all that evolution has endowed us with in the speech department, we are sloppy, even lazy, when it comes to expressing ourselves. We say one thing meaning another, then attempt to clear it all up with more words. We let our minds drift like wayward hot air balloons while we rattle on. "What was I talking about?" we mumble to each other. We drag the great emotional weight of our past and our fears for the future into the simplest of verbal exchanges, like what to get for takeout. We say one thing while our body language or volume broadcasts the exact opposite. "I'm *sorry*," countless spouses through the ages have screamed at each other, to which countless spouses have responded with a bit of body language—slamming a door. Speech can be such a crutch that we focus on what somebody says when what they're doing says it all. As expert trainer Karen Pryor points out in *Don't Shoot the Dog!*, when human couples argue, what they say steals the show rather than what they are doing—fighting.

Humans are so sloppy, I think, because we can later explain ourselves, put it another way, or apologize to our fellow higher primates. There's no explaining anything to an animal. If a trainer's timing is off and he unintentionally teaches a dolphin to jump when he meant it to flip, there's no explaining to the marine mammal, "Oh, jeez, sorry, what I meant was . . ." If a trainer unnerves an animal by getting too close too fast, he doesn't get to explain that he just wants to be friends. When a trainer falls down in front of a big cat, he doesn't get to explain it was an accident, that he's not a prey animal.

That animals take the world literally, connect the behavioral dots on the spot, and respond so clearly, drives home this fact: What you *do* is communication. If it wasn't so, we couldn't train animals. But we can, and without one word.

———

Whether to use training with my husband, friends, and family was really never a philosophical question for me. It was a practical one. I found myself having the same irksome conversations, running into the same old annoyances, over and over with friends and relatives. My mother and I had had the same discussion about her hearing, or seeming lack of it, for years. I could not convince her that she needed hearing aids. Nothing I said ever persuaded my husband to quit losing his temper. I was tired of repeating myself. I needed to try something new. When I saw what trainers could accomplish using basic behavioral principles, I saw that something new.

I once spent an afternoon talking to the debonair Hollywood trainer Hubert Wells, who has worked on a long list of movies from the first *Doctor Dolittle* to *Out of Africa*. He described himself as a "good all-around mammal man," though he never got on with bears. Wells wore a crisply pressed safari suit of khaki. He had retired and sold his company to another trainer, but he still lived on the edge of his old animal compound in a narrow canyon north of Los Angeles. I heard an elephant trumpet or two as we talked. We drank coffee in his dining room while Starbuck, his unruly Jack Russell terrier puppy, lapped the table. "He's like double espresso," Wells quipped. Around us were the photos of Wells's long career, from a black-and-white picture of a vizsla, the daughter of the dog he sneaked out of his native Hungary when he fled the Communists in 1957, to a picture of him cheek to mane with a favorite lion. I asked Wells his general approach to working with animals. He answered in his lush Hungarian accent, "If something doesn't work, try to think of something else."

I took his advice to heart.

"TO SHAMU"

As I describe a problem I'm having with a student in my class who consistently turns in papers late, Scott responds, "Is there a way you can shamu it?"

Shamu the proper noun has become an all-purpose verb in our house. It's shorthand for using the principles of animal training to solve a behavioral riddle. We shamu friends, family, and neighbors. We shamu each other. "Did you just shamu me?" one of us will ask the other. Even a couple of our friends have begun to shamu, and call it that. In fact, it was our friend Kirsten, a high school teacher, who first conjugated the word. She shamued us into using it.

Shamu the noun has also become an adjective, as in "I just had a profoundly unshamu moment." That means I just blew it, as in the time I snapped at the driver of the recycling truck for blocking my car in the driveway. Or we might shake our heads over someone's hysterical behavior and mutter, "That's not very shamu."

THREE

THE ZEN OF ANIMAL TRAINING

pull my car flush with a drive-up ATM, unroll my window, and go to work. First I have to yank a deposit envelope with all my might out of the jammed dispenser. Then I press it against the steering wheel to fill it out. Penny Jane, seeing we might be stopped for a while, curls into her nap position on the backseat. Dixie, who doesn't want to miss one nanosecond of life, stares out the backseat window. The pen runs dry, so I root through the glove compartment for another. Just as I lay my hand on a pencil, which won't do, a loud sigh heaves my way. In my rearview mirror, I spy a young woman on foot behind my car. She tosses her long hair over her shoulder, stamps a high-heeled shoe. I hurry, scrawling my name on the backs of four checks as fast as I can. As I slip the deposit envelope into the machine's maw, another epic rush of breath sounds. Dixie cranes her head to see what the fuss is. The impatient missy bugs her eyes.

"You could use the ATM inside the bank," I call to her as matter-of-factly as I can.

"I'm allowed to use this line," she shoots back.

"Well, it might be *faster* for you to use the one inside." I can't resist letting my voice rise a little.

"Why don't *you*?" she huffs.

"Because I'm in a *car*, you *stoop*—" I stop myself. I am neither thinking nor acting like an animal trainer.

Before I could try any training techniques on my fellow species, that's what I had to learn to do. That did not mean dressing in excessive amounts of khaki or squeezing into a wet suit. If only it was that easy.

Animal training is a mind-bender. Like the Marines, it's not only a job but a way of life. When I interviewed trainers, I heard the same line over and over: "This is not what I do but who I am." Normally I would have cocked a journalist's skeptical eyebrow at that, but I knew they weren't exaggerating.

As one trainer put it to me, training animals "tweaks your head psychologically." It's like a Zen Buddhist lesson in self-control, not to mention an unending mental balancing act of opposites. You must be fast on your feet because all animals are fundamentally unpredictable, yet calm and consistent lest you train something you'd rather not, or, worse, get hurt. You must be responsible, as trainers say it's never the animal's fault, but egoless, never taking personally anything an animal does. You must live in the moment but think ahead, anticipating an animal's next move or response, what one trainer called "proactive second-guessing." You must be confident, so a big cat doesn't think you are prey, but not so confident that you casually bend over and tie a shoe, which would make any big predator consider you his next meal. You must be forever cautious. I once watched a novice elephant trainer absentmindedly walk through a gate with a pachyderm at his side. The senior trainer pointed out that the elephant could easily have crushed him against the gate—just by accident. One second of daydreaming, and squish, next thing you know you are a stain.

Trainers must be self-conscious, always aware of what they say with their own body language. Fluttering hands or squared shoulders can be important bits of information to the animal mind. Given what a fidgety species we are, as if we all suffer from a touch of Tourette's syndrome, that is no easy task. But if you want to train animals, you must overcome your human tics. A six-foot-four teenage student at the training school had to make himself less scary to Rosie the baboon. He learned not to make sudden movements nor laugh in the baboon's presence, which was especially hard as the guy was a cut-up. For once his slouch came in handy, because good posture was out. He had to slouch all he could to look smaller and less threatening. He couldn't look directly at Rosie, which is baboon for "I'm going to get you." He had to watch her, an animal with sharp teeth and superhero reflexes, out of the corner of his eye.

While being ultra-self-aware, trainers must get out of their own skin and consider the world from the animal's point of view. Every time a trainer at the school walked the wolf, she watched for the smallest thing that might trip the canine's hunting drive, from bunnies bounding across the path to a neighborhood Weimaraner on a leash walking just outside the compound. One afternoon I followed a trainer who was teaching three students to walk a youngish male cougar on a chain. We paused and stood in the sun gossiping as the cougar lounged on bark mulch at our feet. The students were practicing how to be relaxed yet alert around a predator. The cat was certainly relaxed. He appeared to doze except for the occasional flick of his tail. Suddenly the staff member stuffed a treat in the cat's mouth. She had heard the distant buzz of the zoo van coming our way, and knowing that cars scared the cougar, wanted to distract him. If frightened, he might run for it, dragging the student behind him. Neither I nor the students had heard the engine. Sure enough, the van passed by, but the cat, crunching chicken necks, hardly flinched.

Trainers must also learn to do the opposite of what their instincts tell them to. Should a killer whale grab you in his mouth, go limp. That way he will grow bored and let go. If a big cat attacks, roll into a ball and don't struggle. Struggling will make the cat bite harder. If a snake sinks its fangs into your hand, don't jerk away. Its teeth, which angle inward, can easily break off in your flesh, where they will fester. A snake bite won't kill you, but the infection might.

Obviously, trainers must have the self-control of monks, because it is how they behave that will make or break them, figuratively and literally. Student trainers are surprised at how much time they have to spend on their *own* behavior. They learn to check their fears when the emu is feisty, step close to the giant bird and hold her by her slender, strong blue neck so that she cannot whack them with her sizable beak. They have to remember not to run near the primates' cages lest they upset the tetchy monkeys. Students can't lose their temper during a training session, ideally not even sigh, because that might undo all they have accomplished to that point, especially with animals as skittish as a Patagonian cavy. Students have to learn not to jump, even wince—exactly what their nervous systems demand—when Kiara the clumsy lioness roars in their faces. Should her roar have an effect, the lioness would be sure to turn up the volume at the same novice trainer again. And so students new to the lioness learn to stand impassive as a statue as Kiara's roar rushes through them.

I needed most, if not all, of the same qualities of a trainer to work with human animals. Though I will never have the composure of a monk, I had found, working with my dogs, wells of self-control and patience. Couldn't I do the same with the humans in my life? Especially given that I, unlike a professional trainer, needn't ever worry about anybody biting me or pinning me to the bottom of the pool, though Scott, like many humans, does roar at times. He cannot, however, hold a candle to a lioness. Having heard Kiara at close range, I found my husband's outbursts more tolerable, if not comparatively tame. The lioness desensitized me to my husband's occasional growl, for which I owe Kiara thanks.

The primary lesson trainers taught me was this: I had to look in the mirror. That's what thinking like an exotic animal trainer ultimately meant. I had to change myself. The onus, like it or not, was on me.

I began to analyze my own behavior. I considered how my actions might contribute to my husband's, my sister's, or those of the huffy young thing at the ATM. What could I *do* differently, I asked myself? Typically, I would have concentrated exclusively on what I should say to someone. Like a diplomat, or a writer for that matter, I would ponder the exact word choice to ask my husband to unpack his suitcase, which languished on the floor long after we had returned from a weekend trip. Or parse out a sentence to explain to a neighbor why I didn't appreciate his keeping his stuff in my backyard.

It's not that I wouldn't use words anymore, but they wouldn't be *all* I used. Instead of obsessing over the right thing to say, I would also consider how to say it, my timing, and my accompanying body language. I adopted the trainers' motto: "It's never the animal's fault." That's easier to apply when the animal is not a human animal, because sometimes it *is* the human animal's fault; but it forced me to think of how I could use my behavior, the *only* behavior I can control. This mental shift demanded a lot of me, but suddenly, I had a new tool to work with—me.

DON'T TAKE IT PERSONALLY

Humans, despite our big, wrinkly brains, are a rather myopic, self-absorbed lot. Maybe if we were prey animals at the bottom of the food chain rather than at the tippy-top, we might be more empathetic. As it is, we so see the world from our limited point of view that we assume the rest of the animal kingdom does too. So we anthropomorphize ad infinitum, projecting all kinds of human characteristics, motivations, and talents on all things furred, feathered, and scaled. We think dogs chew our new pair of Cole Haans out of spite. They don't. We see a dolphin's smiley gape and think they are friendly, but they'll bite you, not to mention ram you, even ram you and then bite you. Remember, everything with a mouth bites. We assume any animal in captivity would like its freedom. We don't really know that, never mind that freedom is a human idea. We come up with the rationales of a preacher (a cougar that dares attack humans is quickly dubbed evil) or a Freudian analyst (my parrot bit me because it's jealous of my boyfriend) to explain the way animals behave. As expert bird trainer Steve Martin has written, in that case, the parrot would bite the boyfriend. Parrots don't displace, rationalize, or sublimate.

We touchy-feely *Homo sapiens* assume all creatures great and small would like a pat on the head, a hug, even a big kiss. Just because we'd love to throw our arms around an orangutan, a sea lion, or a baby panda doesn't mean the orangutan, sea lion, or baby panda is going to like it or even abide it. For many animals, to be embraced is to be devoured. Animals can learn to be touched by humans, but that's the point: They have to learn a behavior we take for granted. All the dolphins in the interaction programs at SeaWorld have been trained to be pet-ted. Even then, touching is for them less a lovefest and more a regimented affair, like meeting the Queen of England. It has to be. When I got to hug my first and only dolphin, I had to carefully follow the trainer's instructions. He told me to touch her only from her blowhole back, and to keep my hands well away from her eyes. Then, as the trainer instructed, I got into the chilly, salty pool and knelt on one knee. He cued the dolphin to swim over my thigh until she rested lightly on my leg. Then the trainer cued me to wrap my arms around the animal. I did so gingerly, oddly embarrassed at my humanness, my need to bear-hug this svelte gray animal. The thing about dolphins is they can't hug you back, but she cocked a black eye at me. I wasn't healed, nor did I have an epiphany other than that dolphins feel smooth and tight like inner tubes. Still, I smiled dumbly, the way you do during a once-in-a-lifetime experience. The dolphin held still and waited for a many-times-a-day experience—getting a fish.

Trainers know that while people are animals too, animals aren't people. Trainers must overcome this deeply entrenched human self-preoccupation, this natural reflex to anthropomorphize nearly anything that moves. They do so for very practical reasons. Projecting human feelings and characteristics onto an animal can lead to bad training decisions. A trainer may think an animal won't do a behavior because it's bored, when in fact the animal just doesn't understand what it's supposed to do, or physiologically can't. If a trainer thinks an animal "likes" him, he may take risks he shouldn't. Trainers avoid even labeling behaviors "bad" or "good." They know there are good reasons for the worst behaviors, even for an attack.

Even labeling animals smart or stupid is anthropomorphizing. Whether they be whizzes or dolts is based on *our* narrow idea of intelligence, and can predispose trainers to expect too much or too little from a particular individual or species. As more and more zoos, with their encyclopedic collections, have hired trainers, species no one had ever thought to train have been trained: dart frogs, spotted eagle rays, rhinos, alligators. The irony is that intelligence has nothing to do with how trainable an animal is. As training guru Karen Pryor has put it, the principles of progressive training work on anything from guppies to Harvard Ph.D.'s.

If I was going to think like an animal trainer, then I needed to stop anthropomorphizing too, only I needed to stop anthropomorphizing humans. This meant not taking others' actions so personally, especially my husband's. Previously, a pile of his stinky bike clothes on the floor had been an affront to me, a symbol of how Scott didn't care enough about me. Now, I considered Scott's behavior and that of others with a much cooler head. Good trainers think of behavior as just that—behavior. They don't think about who's right or wrong, who cares or doesn't, who's smarter or better looking. They avoid *shoulds,* stated or implied, as in "The animal should want to sit and stay for me." They don't project their own motivations, moods, or neuroses on the animal.

I began to see behavior as just behavior too, like the workings of a clock rather than a reflection of my own image or some mini morality play. I gave up *should*s, as in "Scott should want to pick up his clothes," or, as Jennifer Aniston's character in *The Break-up* demanded, "I want you to *want* to do the dishes." I detached in a good way. Instead of asking why Scott would do something *to me,* now I asked why he would do something, period, no me in the equation. I came up with new answers, some so obvious as to be surprising.

Scott left his bike clothes on the bathroom floor not because he didn't love me but because it was, simply, convenient. He figured he'd put them away later, after his shower, but then often forgot. He has a bad memory and a worse sense of smell. It dawned on me that my husband could hardly detect the stench, which to me filled the house like fumes from a volcanic eruption. I'm the one with the sensitive snout.

Why was he often late picking me up at bus stations and airports? If the Tour de France is on television, that is why, but it's typically because he doesn't have the greatest sense of time, never has. He's forever shocked at what the clock reads. No matter how he loves me, that love cannot override this deeply ingrained trait. I can't say that it made being the last one waiting on the curb in front of the airport scanning the approaching cars for our Jetta wagon any less annoying, but I no longer think of it as a litmus test of Scott's devotion to me.

Not taking your spouse's actions personally is liberating but no easy task. Mates have a big effect on each other's lives, and, obviously, some behavior *is* meant as a slight and *should* be taken personally. But I realized that I, like many mates, took way too much personally, that I saw offenses where none were intended. Sure, in an ideal world Scott would express his deep, abiding love for me in his every action, even in his motivations. He would even *want* to pick up his bike clothes, because making me happy would be his raison d'être. And in AmyWorld, I could also fly, my dogs could talk, merry crowds would chant my name, and expensive designer shoes would wash up on the beach near my house for the taking. A gal can dream. Or she can think like an animal trainer.

———

I turned a more neutral eye to the other people in my life. Why did my mom get so angry whenever I brought up getting hearing aids? Because it usually made me back down, meaning it worked. Moreover, the idea of getting hearing aids screamed "Old" to her, which rattled her still-young soul. She was mad, not so much at me, but at life.

Why was a close pal of mine so hard to make plans with in advance, say, a dinner out or a shopping trip a couple of weeks down the road? Not because she didn't value our friendship, I realized, but because I, with no kids and a flexible work schedule, was one of her few friends who could often do something at the last minute. Between her job and her two children, my friend, like a lot of middle-aged people, has a heavily scheduled life. I'm one of the few spontaneous elements in it. She liked keeping me that way, I think.

I realized that I took even my own behavior personally, that I saw my actions as proof of my gaping personality flaws, when I shouldn't. So I stopped anthropomorphizing myself and considered afresh why I rush so through the day. Not because I'm a raving maniac, as I'd always assumed, but simply because two professions, waitressing and newspapering, have trained me not to waste one second. Restaurant work made me into such a multitasker in constant motion that I can't walk from the living room to the kitchen without turning lights on or off, collecting messy stacks of newspapers and stray coffee cups, straightening the tablecloth as I pass the dining room table, and making a mental note to call a window washer this spring, all at a clip that makes my husband dizzy.

And finally, why did that young woman sigh with both lungs behind me? Not because she was a bitch, but because she was in a hurry and wanted me to step on it. This aggressive display had probably worked for her hundreds of times before. She was a peacock flashing its tail feathers, a chimp with its fur on end, a wolf baring its teeth. The huffing and puffing was a way for her to vent frustration, to let loose some of that adrenaline-fired angst, to win a fight without actually fighting. Okay, and maybe she was a bitch. Regardless, I could have been anyone. She was not after me, Amy Sutherland, but the person keeping her from the ATM machine.

So why did I stop myself in the middle of loading both verbal barrels, especially when I'm such a good shot? Because I realized I was responding to the whole pesky exchange emotionally, personally, which was just one way I wasn't thinking like an animal trainer. I was also roaring back at her roar. I was responding to what she was saying when I should have ignored what she was doing: picking a fight. She was the monkey grabbing my ponytail, and rather than resist my instincts, I was running with them, pulling back with all my might, making it so much more fun for her. My self-control was obviously shot. Time for me to leave the cage, so to speak.

I took a deep but silent breath of my own, shut my mouth, grabbed my receipt, and put it in my purse. As I pulled away without looking back, I heard the young animal behind me let loose with one last exaggerated sigh. In my car, where I knew she couldn't see me, I responded—with a smile.

SET YOUR ANIMAL UP FOR SUCCESS

At the training school, students didn't train Sequoia the mule deer when the east winds raked the teaching zoo. The demure deer hated the hot gusts, so she jumped against her enclosure when they blew. Students never handled the big snakes while they molted. When snakes shed, the skin over their eyes loosens and clouds their vision. That can make a boa or anaconda jumpy, especially when some out-of-focus hand grabs it. A molting snake is more likely to bite you, plus the unnerving experience, from the reptile's point of view, that is, might color the snake's opinion of being handled. Precious the yellow anaconda might be harder to pick up next time, whether she's shedding or not.

Trainers want their animals to do well. As they say, "Set your animal up for success." They do this in a number of ways, from setting realistic expectations (keeps the animal from getting frustrated) to keeping training sessions short (keeps the animal from becoming bored). That's also why they do not train an animal having an off day for whatever reason. A new mother camel is in no mood to sit and stay. A sick sea lion isn't up to standing on its tail. A beluga whale bullied by another whale won't be able to concentrate on a hand signal. A trainer not only wastes her time with an out-of-sorts animal, but she can also make training a bad experience. This is yet another reason a trainer must be an astute observer. She must also be self-aware, because setting your animal up for success goes both ways. Both animals, the trainer and the trainee, need to be set up for success. To that end, trainers don't train when they themselves are sick, distracted, grouchy, or just emotional. They won't be on their toes. Their timing will likely be off, or worse, they might miss a subtle bit of body language broadcasting, "I'm going to take a swipe at you."

What if the animal and the trainer are both off? One summer evening in 1928, Mabel Stark, the then world-famous tiger trainer, stepped into the cage with her big cats for a circus performance in Bangor, Maine. In short order, two male cats, Sheik and Zoo, each

41

bit chunks out of the petite trainer in her white leather jumpsuit. That was just for starters. Stark spent the next two years in and out of hospitals recovering. The tigers, it turned out, had not been served dinner and had spent the day lounging on wet hay. What of Stark? The tigers had not been fed because the circus had arrived late to town. For the same reason, Stark may have been rushed, not an ideal state of mind for even the best tiger trainer.

Like Stark, so many of us figuratively go into the cage when we shouldn't. Can you reason with a tired, hungry, or upset kid? Not from what I've seen. Can you reason with a tired, hungry, sick, upset, or hungover friend, spouse, parent, employee, or boss? Not from my experience, but so many people persist. We often pick the worst moment, say, when someone is frantic over a lost pet, purse, or paycheck, to drive home a point ("If you just used a leash, or kept track of your stuff, or deposited the check like I told you, this wouldn't have happened!"). We may mean well, but that point typically falls on deaf ears and can provoke a swipe. People, like animals, aren't wired to learn lessons when they are out of sorts. I know I'm not. When my asthma flares the least little bit, all I can think of are the uncooperative sacs of air otherwise known as my lungs. When my blood sugar wanes, all information that doesn't have to do with grilled cheese sandwiches begins to bounce off me. I'm slow to lose my temper, but once I do, I'm stone deaf to even the most diplomatic suggestion. Advice at that point works as well on me as it would on a charging rhino. Best to shut it and jump behind the nearest tree. By the same token, I don't try to "train" when I'm cranky, sad, or hurried. I'm in no state of mind to control my behavior, most specifically my tongue. If the odds of having any kind of productive interaction are low, I should not go in the cage.

Like a trainer, I began to pick my moments better. Just now is not one. Scott and I are late yet again. Neither of us is particularly prompt, and so we speed through the dark toward a dinner party. Scott curses the clock and floors it as the traffic light flashes to red. As we sail under another "long yellow," I clench the wooden salad bowl in my lap. I've been in my share of car accidents, one of which landed me in the hospital and left a tidy but long scar running along my spine. I can get nervous on four wheels, especially when Scott rushes. Still, this time I don't say a word.

Asking Scott to slow down just now would be like picking up a shedding snake. And I might as well be molting too. Instead, I double-check that my seat belt is fastened and go to my happy place, the T.J.Maxx in my head.

FOUR

KNOW YOUR SPECIES

Consider *Elephas maximus* from a trainer's point of view. It matters, for example, that the Asian elephant is a herd animal, prefers company, and responds to hierarchy. Their wondrous trunks are strong enough to lift a large log, yet dexterous enough to pluck a penny from the ground. With their great bulk, elephants cannot jump, but they can stand on their hind legs or heads. Ravenous herbivores, they are round-the-clock noshers, devouring up to two hundred pounds of bark, twigs, leaves, seedpods, fruit, and grass a day. Thus they are epic poopers. They live in a state of grace few species are blessed with. They are essentially neither prey nor predator.

These trivia have very practical implications for a trainer. To start, there's no point in trying to teach the world's largest land mammal to jump. However, given their powerful yet flexible proboscis, you can train an elephant to gently lift a human. That they are such chowhounds means they respond eagerly to rewards, especially bananas, watermelon chunks, and palm fronds. That they poop so prodigiously means you'll spend more time scooping than training. That they aren't predators means they won't ever see you as dinner, as a lion might, but given their size, they can still easily kill you.

The best animal trainers learn all they can about a species: its natural history, feeding habits, anatomy, social structure, and native habitat. They want to know what might spook it, if it's up and at 'em at 6 A.M. or 6 P.M., whether it prefers privacy, if it hunts for dinner or is dinner or both.

They bone up on the ins and outs of a species because all this information makes a difference when it comes to training. You don't want to train a winter animal—say, an Arctic fox— in the sun. He'll melt before you can teach him to shake your hand. You can't expect much from a nocturnal animal at high noon. You'll spend most of your energy just waking the sleepy-head up. Good luck teaching a walrus to fetch. It's inclined to swallow anything in its mouth. You don't want to train a mule deer, a prey animal, near the Bengal tiger, a predator. The deer won't be able to concentrate on you with such an impressive pair of teeth nearby, even if the teeth are in a cage.

Though modern trainers use the same basic principles to work with all animals, they can't train one exactly like another. A cookie-cutter approach would fail because Mother Nature likes her diversity, and so the animal kingdom brims with individuality. As the legendary Hollywood trainer Hubert Wells put it, "From species to species no two animals are alike, not even two insects."

For example, a spinner dolphin, unlike its cousin the bottlenose, is not keen on toys nor playing nor human touch. River otters explore with their nose; sea otters, expert foragers, with their feet. Better to train a river otter to touch a ball with its snout, and the sea otter to do so with a paw. African elephants are generally considered more skittish and aggressive, not to mention they have bigger tusks, than the comparatively docile Asian elephant. The African can be trained, but the Asians are more willing, not to mention safer, students, notably the females.

A good trainer also learns all he can about the individual animal, its own spin on its species: where it likes to nap, its best friends and favorite toys, its general health, its life story. At the training school, the two olive baboons had vastly different personalities, given two very different backgrounds. Olive, Miss Nervous, was caught in the wild as a baby, then raised in solitude, which is never good for a young social animal. Consequently, she had many neurotic habits, including playing with her feet as if they were puppets and, when she was upset, notably when students cleaned near her cage, shaking herself wildly, even banging her head on her den box. Rosie, Miss Confidence, was hand-raised in captivity. She is quite at home with humans and is, in general, a happy camper. She often lounges on her back on a shelf in her enclosure and rests her head on a crooked arm. When student trainers worked with Olive, they focused on soothing her, building up her nerve. When student trainers worked with Rosie, they focused on behaviors, like teaching her to ring a bell or jump into their arms.

Details such as size, gender, and age matter as well. Young

animals might not have a long enough attention span for some training. Old animals might not have the physical prowess. Schmoo, the dowager sea lion at the school, was still scarily sharp at a creaky twenty-three, but she had cataracts, which meant she couldn't see hand cues clearly. The student trainers relied on verbal cues. Then Schmoo's hearing started to go. Sometimes the students had to turn off her gurgling pool filter so the sea lion could hear them.

THE SPECIES IN MY LIFE

Inspired, I turned a trainer's eye to the lovable but sometimes baffling species known as *American Husband*. They are hierarchical and territorial animals, especially in matters concerning who holds the remote control or adjusts the bass settings on the stereo receiver. They are keen-sighted but have neither night vision nor refrigerator-light vision. They are unable to hear higher decibels, especially those that make up spousal speech. Many eat a largely carnivorous diet, which includes cows, pigs, birds, bird eggs, and some grains, preferably fermented or in chip form. Some hibernate in the cold season, otherwise known as the football/basketball/hockey season. Some also hibernate in the hot season, otherwise known as the baseball season. They can be crack tool users, though many are satisfied just to have the tools.

The subspecies known as Scott is a loner, but an alpha male. So hierarchy matters, but being in a group doesn't so much. He has the balance and grace of a lemur but moves slowly, like a tree sloth, when getting ready to go out. Skiing comes naturally, but being on time does not. He's diurnal but can easily become nocturnal as a jaguar, especially when he discovers a new trove of Borat videos on YouTube. His native habitat is the crisp climes of Minnesota, where the snow is plentiful and conversation is not. At six feet, he is bigger than typical for his species, not to mention smarter. However, his memory is unpredictable. He can beat most anyone in Trivial Pursuit, but only because the game does not include questions like "Where is your wallet?" He loves snakes but cringes at spiders, pretty much his only fear. He needs exercise, so much that if he doesn't get to pedal his bike regularly or is cooped up inside an office for days on end, he can become cage-aggressive. He's an omnivore and what a trainer would call food-driven. However, like a big cat, he can subsist off one enormous meal, say a stack of buttermilk pancakes made by yours truly and doused with maple syrup, all day if need be. His favorite resting place—the bathroom, reading. His cage-mate—me.

What of the other human animals in my life? My mom is a social creature but doesn't give a hoot about her rank. She will always go along with the herd, even if the herd is headed up a steep, boulder-strewn trail, even if she is in her seventies and afraid of heights, even if her daughter has forgotten the trail map. She is good-natured but guards her food, especially salted peanuts, and becomes aggressive when you call her a hillbilly or suggest getting hearing aids. She is as omnivorous as a possum (she once ate a leaf off a plant in a doctor's office) and part scavenger. I came to after back surgery to the sight of her eating my hospital dinner: "This is pretty tasty," she reported as she shoveled in a forkful of limp green beans. She is diurnal but flirts with being nocturnal, which typically results in her snoring loudly during the eleven o'clock news. Unlike most humans her age, she still craves novelty and activity, which makes her behavior super plastic, as a trainer would say. She's also freakishly strong and doesn't think twice about lifting a fifty-pound bag of mulch. Other than heights, nothing much scares her, except the prospect of having nothing fun to do.

In the species known as *Younger Brother*, there is the subspecies Andy. He is typical of his species in that he has remained very playful into maturity, which he expresses by asking me to pull his finger or belching loudly while I'm on the phone doing a live interview with a radio station. At six feet five inches and 240 pounds, Andy is big for his species and so, like an elephant, can easily unintentionally hurt you, say, when he plows into you while demonstrating how he can ski on one ski, almost. Also like an elephant, he is powerful. He has a black belt in karate, can take down a wall of plaster, lath included, in minutes, and, most importantly for me back in my apartment-dwelling phase, can hoist a room air conditioner onto his shoulder. He is not afraid of any stimuli that I can tell. He is food-driven, especially given that food, epic amounts of it, is needed to produce the gas that is key to much of his playfulness. He wilts in the heat and thus migrated away from his home range of Cincinnati to Denver the first chance he got. There in the dry air and sunshine he thrives, and seems to have somehow grown a few more inches.

In the gregarious species known as *Girlfriend*, there are subspecies Dana, Hannah, Becky, Nancy, and Elise, among others. Typical of the species, all are highly communicative animals, notably Dana, whose native habitat is Ohio, where conversation and friendship, like fruit on the jungle floor, can be found easily. They also have keen eyesight, and most are easily distracted by shiny objects, especially those hanging from other women's earlobes. Nancy, like a peregrine falcon, can spot a designer cashmere sweater marked half off from afar and swoop down on it at fantastic speed. Still, they all have long attention spans, whether it be for a lengthy and heartbreaking story about how I spent four years knitting a sweater on itsy-bitsy needles only to have it not fit, or detailed discussions about evolution, Italian Renaissance painting, or the ethical implications of stem cell research. None are small for their species. In fact, most are tall, which might explain their general aplomb. They do not spook easily. The diets of all include wine, coffee, chocolate, and medium-rare steaks. Hannah cannot digest dairy or wheat. Becky can, and with relish. Elise's feeding habits—no breakfast, maybe no lunch, and one big early dinner—make her somewhat crepuscular, like a caracal, that wild cat with the fancy tassels hanging from its ears who hunts for birds come twilight.

If I am to think like an animal trainer, I must also know myself. So what of subspecies me? I am intensely diurnal and active. Consequently, like a nectar feeder, I need a steady supply of calories throughout the day. Without fuel, I crumble. Frightening stimuli include heights, hypodermic needles, and dark basements. Regardless, I'm what a trainer would consider a confident animal, not to mention a curious one. In fact, too much routine makes me pace. Like my mother, I require novelty. Like a sea otter, I can forage all the livelong day, digging with my paws through clearance clothing racks, housewares stores, even my own closet. I have a bad dust allergy, one of the few things that can curtail my foraging, especially in antiques stores. My favorite resting place—the beach. I am gregarious but need solitude to read, think, and dream.

Walking down a busy city street, I now see my fellow higher primates—*Homo sapiens*. We, males and females, are an intensely hierarchical species, which we express in millions of little and big ways every day, from the tone of voice we use with each other to which country gets to sit where at the United Nations. We are also territorial and plant our flags readily, whether with a picnic blanket or a chain-link fence. Collaborative yet competitive, we are a species continually at cross purposes. We have sharp vision, serviceable hearing, and a substandard sense of smell. We prefer privacy for sleeping and mating, but not for eating, even the food guarders among us. We are such omnivores we can afford to be picky eaters, and even express our place on the social ladder through our choice of entrée. Our native habitat is pretty much anywhere except the moon (so far). We are fundamentally bipedal and terrestrial, but we climb trees, swim, and, thanks to our opposable thumbs and tool use, fly. We have a highly developed brain capacity, but that is the key word—capacity. Physiologically, we can stand on our heads, cartwheel, and flip, but most of us would rather not. The younger members of our species, as with all animals, are energetic, even reckless, especially the males, which can make them annoying to us more mature members. We are long-lived, in large part because we, even more than elephants, are not preyed upon, except for the occasional shark or grizzly bear attack. Consequently, we are generally self-assured, which gives our curiosity free rein, but also makes us careless. We are highly communicative, so much so that we, like parrots, are happy to communicate for communication's sake alone.

———

As funny as it is to consider my fellow humans this way, there is a point. If I want to set my animal up for success, I need to know what comes easily to them and what doesn't. For example, Scott, like a nocturnal animal roused too soon, wakes each morning as if he's returned from the dead. Early morning flights or early morning anything is a trial for my husband. Getting him up before 7 A.M. is like teaching a walrus to fetch. It can be done, but a lot of toys are going to be swallowed along the way.

Better to play to a species' strong suit. Birds of prey can be taught to hunt on command because they are natural hunters. Baboons jump high on the African grasslands, so it's no stretch to teach them to do so on command. The same goes for wild dolphins, which flip and arc like Olympic gymnasts over the open seas. Likewise, Scott, being a Minnesotan, is given to all things snow: skiing on it, looking at it, talking about it, and, most important to me, shoveling it. All I need do is hand him a shovel and the flakes fly. He likes clothes, furniture, and eating lunch out, so, unlike many husbands, he readily makes shopping trips with me. Given his nocturnal tendencies, he can stay up all night to finish painting a room in a pinch. He's flexible and curious, which makes him a natural traveler, and fearless, which gives him the nerve to steer a rental car through a foreign city.

Instinctive behaviors can cut both ways, making training

easier or harder. Then they just are a fact of life. Trainers know that some behaviors are so hardwired, they may never be trained away. Good luck stopping a camel from spitting or a homing pigeon from homing. A fox was born to burrow and a parrot to preen. Even the best trainers will have a tough time convincing either animal otherwise. In fact, doing so may be a big waste of their time.

Like a trainer, I accepted that some behaviors may be too entrenched to change. Just as you can't stop a badger from digging, there's no stopping my husband from losing his wallet and keys. As geese migrate, so will my husband spend a chunk of each July before the TV watching the Tour de France. For three weeks in midsummer, the sound of announcer Phil Liggett's crisp British consonants and the clicking of so many bike derailleurs turning and turning fills our house. Likewise, my man is a bathroom reader as a polar bear is a carnivore. I can remove all the bike catalogs and hide the pop music books, but next thing I know he's in there with a weighty Don DeLillo novel. It would be easier to remove a spider from a room myself than to persuade Scott to do so. However, if you need someone to pick up a snake, he's your man.

PUNISHMENT IS IN THE EYE OF THE RECEIVER

In getting to know their animals, trainers learn what the animals like and what they don't. The latter can be as idiosyncratic as the former. For some social animals, the trainer's leaving at the end of a session can be a negative. I met some spider monkeys who couldn't stand to have their trainer leave the cage. He would have to uncoil their tails from his arms and peel their black fingers off his neck. For other animals, the trainer's presence, especially if he moves close, can be punitive. That was the case for a cavy, a South American rodent, at the training school, who was so scared by people near his cage he would jump against the wire in a desperate attempt to escape.

What gave me food for thought was how trainers look to their animals to decide what qualifies as punishment and what does not. If a toy that seems totally innocuous to the trainer makes a monkey nervous, then it's punishment for that animal. It's not whether the trainer thinks the toy is scary but whether the animal does. If turning your back to a dog during a training session makes her cringe, that is punishment, even if it seems meaningless to you.

We humans typically take the opposite approach with each other. What qualifies as punishment, its scale and method, is typically decided by the punisher. That approach produces some ineffective, out-of-proportion, and even accidental penance. That's because people tend to reprimand each other in ways that would be punishing to themselves. If you hate to be yelled at, chances are you'll yell at someone to set them straight. But not everyone hates to be shouted at or hates it as much. A select few love a good screaming match. Along those lines, if you don't mind roaring, you might not realize how punishing it is for someone else. You might let loose with both barrels thinking it's not that big a deal, but the person you're yelling at might burst into tears. If you think like an animal trainer, you'll note the tears and hush. Well, if you were a progressive animal trainer, you wouldn't even be yelling.

I developed general rules for behavior I labeled "instinctive," which didn't mean it absolutely couldn't be changed, but did mean it would be hard going and probably not worth the effort. My first rule had to do with the relative age of the behavior. Generally, the longer a behavior has been around, the harder it is wired. For example, even back when my husband was my boyfriend, he needed a GPS to locate his wallet. That's a good fifteen years of wallet losing right there. My guess is he's been misplacing his wallet since he got his first one in eighth grade.

Next I ask, Does the behavior stem from a fundamental trait? My husband is a bit dreamy, thinky, which makes him a good writer but inclined to go through the motions of daily life without paying attention, such as noting where he set down his wallet as he came into the house. That detail was lost while he hashed out a paragraph in his head or plotted the course for a bike ride later that day. Given that two of my rules applied to Scott's wayward wallet, I gave up any notions of changing that about him. In this game, two strikes and the behavior is out, as in, no training plan for that.

Lastly, if someone is completely unaware they do something, my motto is "Beware." Whistlers, table drummers, change-in-the-pocket jinglers, most don't even know they are making a commotion until someone yells "Cut it out!" We all have these little habits that are so deeply rooted in our subconscious that we are oblivious to them. When people point out these little quirks, we respond, "I do?" My mother uses the word "cute" (as in "Isn't that ka-*yute*?") constantly. When you tease her, she'll respond, "Have I been saying that a lot?" I often forget to zip my computer backpack closed, though I've had strangers, even a Boston police officer, point this out to me. I laugh really loud, which I was unaware of until my husband, embarrassed in a reserved Minnesotan kind of way, took to patting me on the leg each time I guffawed. Despite the patting, I still hoot whenever something cracks me up. I can't help it. My brain sees something funny and shoots a message to my vocal cords—"Hit it!" I also can't help leaving drawers open. You can chart where I've been in the house because I leave a trail of gaping kitchen drawers, dresser drawers, buffet drawers behind me wherever I go. This drives Scott crazy, but try as I may to remember to close the drawers, it simply never occurs to me. My brain says "Open." It does not say "Close." Scott has the reverse problem with curtains. Given that he is a very private subspecies, he pulls the drapes if he is going to take off his watch. Then he leaves the darkened bedroom. I've asked Scott countless times to open the curtains. I once even explained how the Shakers believed naturally lit rooms were essential to good health. He'll say that he *does* open them. He doesn't. The proof is in the dimly lit room.

Now, we could both think up brilliant training strategies to change these behaviors we find annoying in each other, but the jobs would be long and uphill, like trying to train a raccoon to stop washing its food. Just as I want to set my animal up for success, I want to do the same for myself, so best not to set my sights on the next-to-impossible, especially if in the scheme of things the behavior is a relatively small thing. So I don't mention the curtains anymore. Writing this, I realized that Scott doesn't mention the drawers anymore either. It's not worth bugging me, he says. Animal training may finally have taught us how to pick our battles.

Besides, even if you could get a raccoon to stop washing his food, would you want to? That's a big part of what makes a raccoon a raccoon. Trainers don't want automaton animals. A big part of what makes humans humans is that we aren't perfect. We have quirks, tics, little genetic blips that makes us *us*. To know your animal is also to accept it, instincts and all.

FIVE

WHY I STOPPED NAGGING

can smell Scott's bike clothes from my desk. The reeking
pile lies on our bathroom floor, where Scott left it after his
shower. He rode thirty-five miles, he told me proudly. Here at
my desk, I can smell every bead of sweat each of those thirty-
five miles wrung out of my lean husband. I wrinkle my sensitive nose, shut my office door, and get back to work.

In the past when Scott left his fetid bike clothes to ferment
in the bathroom, I'd ask him to pick them up. If he didn't, I'd
ask again and again, pinching my nose tight or clutching my
neck as if I were suffocating. I'd ask him if the EPA had ap-
proved cleanup of the toxic waste site in our bathroom. But
eventually I'd lose my sense of humor and grow more annoyed
with each request.

I repeated myself a lot to my husband, requesting over and over that he dispose of his dirty tissues clustering near the gearshift in the car. I'd call through the bathroom door "Are you almost ready?" several times before we went out to a movie. I started lots of sentences with "I hate to sound like a broken record, but . . ." Not even his spousal deafness, which increased in proportion to how much I repeated myself, stopped me. By the time he did what I asked, I was usually too miffed to manage a thank-you.

I was a nag. Maybe I had a creative delivery at times, but I was a big, fat nag. This made me even madder at Scott. He forced me to whine and complain to get what I wanted. Or so I thought, until I saw how animal trainers do it—or rather don't do it.

I noticed trainers did not get a sea lion to salute by nagging. Nor did they teach a baboon to flip by carping, nor an elephant to paint by pointing out everything the elephant did wrong. In fact, the trainers I followed rarely, if ever, even corrected their animals.

———

Progressive animal trainers reward the behavior they want and, equally importantly, ignore the behavior they don't. This revolutionary approach comes from the world of marine mammal trainers. They didn't entirely invent this method (it was known as "gentling" in the circus world, "affection training" in Hollywood), but they were the first to codify it, show how well it worked, and use it exclusively. You won't find a marine mammal trainer who doesn't work exclusively with positive reinforcement. At least you shouldn't.

The early dolphin trainers had a toothy, flannel-gray puzzle on their hands. How do you train an animal you can't get your hands on? They couldn't rein the marine mammals or leash them. If a dolphin didn't like training, it could swim off or sink below the water's surface. Even though they were in captivity, the dolphins could not be made to train. They had to be enticed. How?

The answer was found in operant conditioning, which was hashed out by psychologist B. F. Skinner at Harvard University in the 1930s. In his lab, Skinner demonstrated that behavior is affected by its consequences. A good result encourages an act, making it likely a living thing will repeat it. If a pigeon pecks a piano key and a seed appears, that bird will play on. Enough well-timed seeds, and you should have a winged Rachmaninoff on your hands. That all creatures will repeat what gets them the goods seems obvious, but Skinner proved this to be scientifically true and to be fundamental to the way all animals, including humans, learn.

Skinner found a negative result can change behavior, too. If the pigeon pecks a piano key and gets a shock, it would seem a pretty safe bet the bird will give up dreams of a concert career. But the behavioral equation gets a little tricky with punishment. Skinner's experiments showed correction had an effect but not a predictable one. The pigeon might just come up with a way to peck the ivory, maybe very lightly, *pianissimo,* and avoid the shock. Or if he is a rare bird, the Glenn Gould of pigeons, he may decide that his art is worth occasional, even regular, pain. Eventually, he may become desensitized and stop noticing the pain at all. He could even come to like the buzz.

DESENSITIZATION

As the very first step of training Kiara the lioness to go into a crate, the students lugged the thing into her enclosure. They left the big wooden box there so she could get used to it. Get used to it the big cat did. The lioness took to lounging on top of her crate, stretching out on her belly, draping her tail lazily down one side. It became her favorite spot to nap.

In this way the students desensitized the cat to the crate. Specifically, they habituated her to it, which is a passive way of desensitizing her. They just let time do the job. The longer the crate was in Kiara's cage, the more familiar it became, and the easier it would be to train the big cat to get in it eventually. By the same token, this is why punishment can lose its punch over time. Animals, humans included, can get used to most anything, and so you find yourself turning up the shock collar and grounding the kid for longer.

Habituation is one of the ways trainers introduce animals to anything new that might unnerve them or to accustom them to something that clearly does unnerve them. For example, trainers might wear crisp white vet shirts to help animals get used to what a vet looks like. Or they might use counterconditioning, an active way to desensitize an animal to something unnerving. When you countercondition, you take a negative experience and make it a positive one by pairing it with something good. The cavy at the school was terrified of humans, which meant if this South American rodent needed any kind of vet care, it had to be caught, which meant chasing the animal around its cage, which only made it more terrified of humans. As a training assignment for a grade, a student set about counterconditioning the cavy to at least one human, herself. Each day she'd inch closer to the cavy's cage, and if the animal didn't hide behind a bush, the student would reward it with a spoonful of alfalfa pellets. Over time, the cavy allowed the student to come closer and closer, until one day the animal ate some pellets right out of the student's palm. A bad experience had been made at least a neutral one.

Humans use habituation all the time. We get used to the sound of traffic over time. We get used to the dropping mercury as the winter wears on and again to the rising mercury as the summer heat peaks. Over thirteen years of marriage, Scott has habituated me to his piles of mail, clothes, and books. Likewise I became habituated to the clamor of a newsroom, so much so that when I first left the paper I had to habituate myself to the quiet of working at home, of no one yelling baseball scores over my head or cussing out their computer nearby. Researching my last book, I spent so much time in the close confines of the reptile room that I habituated myself to one of my deepest fears, snakes. The hours I spent around the snakes, watching students feed Ceylon the easygoing python a baby rat or lug Precious the yellow anaconda into the sun for a stretch on the pavement, did the trick. The more I was near them, the less the reptiles spooked me. I even touched one, a corn snake. It was dry and smooth. I knew I had really lost my fear when I seriously considered buying one of the school's baby sand boas. The only thing that stopped me was the flight home. How does one take a baby snake through Security?

What I personally could use more of is counterconditioning. If someone handed me a candy bar, preferably with caramel and nuts, each time I went a floor higher in a skyscraper, I might get over my fear of heights. If a nurse gave me a pair of earrings each time I got a shot in the arm, I might get over my needle phobia. As it is, some lab techs seem to relish waving the needle in my face. That's like chasing a skittish cavy around its cage.

Human animals, with our love of drama and our innate impatience, like to conquer fears in one huge, dramatic leap. The television show *Fear Factor* is premised on this flawed notion. Afraid of heights? Jump out of a plane with a parachute. Bugs freak you out? Stick your hand in a jar of spiders. Scared of tight, dark places? Go spelunking. The only reward is having survived the experience, and that may be no reward at all if you come out of it twice as scared.

There's a kinder, more productive way to face our fears. Just ask the cavy.

Punishment produces the same unreliable results with humans. There are endless examples. For starters, take speeding tickets. The threat of such discourages many a leadfoot, but others hit the gas while they keep an eye peeled for speed traps. Some people love speed so much that, tickets be damned, they just blithely rocket down the road.

A more specific example can be found in my part-time home of Portland, Maine. When the small harborside city introduced its first recycling program, it did so with a punitive twist. Residents have to purchase certain blue trash bags for regular garbage. Any other kind of bag will be left on the curb. The thinking is that people will recycle more so they will generate less garbage so they can buy fewer trash bags. Some people did. Others refused to buy the bags and stuffed their household detritus into park trash cans, which began to overflow. Some began stealing the blue trash bags off the grocery store shelves. Eventually, stores moved them behind the counter, with the cigarettes. Now you have to ask the cashier for a pack of smokes or city garbage bags.

WHY REWARDS WORK BETTER

Skinner's research showed that positive reinforcement is the surer way to get the behavior you want. Positive reinforcement is how Skinner taught his research pigeons to walk in figure eights, to play Ping-Pong, and even to guide a missile (during World War II, the U.S. Defense Department gave Skinner $25,000 for this purpose but later discontinued Project Pigeon). Progressive trainers today use it too. Not because they are sainted souls vying for the Nobel Peace Prize, but because they are exceedingly practical creatures. They use positive reinforcement not because it is politically correct, though it is, but because it works better.

Positive reinforcement is more effective for a number of reasons, foremost because it is more motivating. Think of it. When do you put in the extra effort? When you might get something you want—some praise, a kiss, a bonus? Or when you are avoiding something you don't want—criticism, a frown, a pay cut? It's a relief not to get kicked in the pants, figuratively or literally, but not much of a reward. That's why all creatures typically do the bare minimum required to avoid a rapping. Punishment may motivate, but it doesn't put a spark in the eye the way a reward can. Take taxes, for example. Do you fill out the forms and file them with any zest? Maybe when you are due a big refund. I doubt it if, like me, you typically owe the feds. Given I'm only filing taxes to avoid a fine, I expend the least amount of energy possible, which means putting off organizing my files and filling out forms to the last minute and dropping them in the mail slot the day they are due with legions of other undermotivated human animals. Call us procrastinators and tsk-tsk all you want, but what do we have to gain by sending them in sooner? I suppose we could brag to everyone how we'd sent our taxes in and then make fun of the dawdlers, but that is not much of a reward to me.

Progressive trainers want nothing less than zest, spark, joie de vivre. All of which could be found at the training school, especially at the hyena's enclosure. A student trainer needed hardly step near Savuti's cage, and he would become a whirling dervish of behaviors. He'd sit, fetch, circle, pick up a log, put his front paws up neatly on the shelf in his cage, all without one cue. He'd look at his trainer with bright eyes as if to say "What about this, or this, or this?!" He was "throwing behaviors," as trainers call it—strutting his stuff to see what might earn him a chicken neck. Savuti was so engaged, the students taught him to create his own moves, what they called innovative training. In response to a hand cue, the hyena would do something he thought up, such as one of my faves, when he put his front paws on the cage, then leaned his chin on them like an ingénue.

One day I watched as Savuti pressed his long, spotted neck against his cage on command. A student then lightly pinched his furry neck. Savuti stayed put as requested. The student was training him to eventually let a vet draw blood from his jugular vein. By exposing his neck like that, Savuti put himself in a terribly vulnerable position, even for a hyena. Why would a hyena do that? Savuti did so because all his experiences with the student, indeed all his experiences of training, had been good—fun, even. He expected more of the same. He trusted her.

This is another big advantage of using positive reinforcement. It builds trust. If a trainer is a source of all things good to an animal, that makes for a pleasant, uninhibited working relationship. Training guru Karen Pryor puts it this way: An animal does something and either gets rewarded, or nothing happens. With no punishment, an animal has nothing to lose, no reason not to trust the trainer.

Steve Martin, who specializes in free-flighted birds but has worked with an encyclopedia of species, has likened this trust to a bank account. Whenever a trainer interacts with an animal positively, he makes a deposit into the good-vibe account. Whenever a trainer does something an animal doesn't like— bops a big cat on the nose, hoses a rhino to make it move into its night quarters, takes a toy away from a monkey—he makes a withdrawal. Best to keep a very high balance, especially, in Martin's case, when the birds can just spread their wings and fly away.

Lastly, with positive reinforcement, you teach an animal what you *want* it to do. You encourage rather than discourage. The problem with discouraging is that it only stops a behavior you don't want. It does not automatically encourage or teach what you want. At the school, student trainers spent a lot of time and energy stopping a camel, the mischievous Kaleb, from tantruming. When he pitched a fit, there would be much yelling and leash jerking. Eventually the camel would calm down. But keeping him from thrashing and kicking was not the same as teaching him what they wanted: to walk nicely on a lead.

We humans assume that pointing out what we don't want makes clear what we do. Parents especially can become so consumed by stopping bad behavior (fighting and shouting), they completely forget to teach kids what they want (sharing and playing quietly).

When I was fourteen, I rear-ended a car with my bike, landed on the trunk, then bounced to the pavement. I was banged up, as was the love of my life, my orange Schwinn boy's ten-speed, but we were both okay. The next night, I got back on the Schwinn that threw me. That ride ended with me walking my bike home again, this time with a broken wrist and a bruise in the shape of my brake line up my thigh. As I limped in the kitchen door with one arm slightly longer than the other, my dad, who rarely disciplined us four kids, took one look at me and barked, "I'm grounding your bike." This was an understandable but rather heartless response. Moreover, it was pointless. He meant well—to stop my run of crashes— but he couldn't ground my ten-speed forever. Before long, I was back in the saddle, despite the cast on my arm. In the interim I had learned exactly nothing about how to pedal more safely, and now I was steering with one usable arm. My father's punishment had taught me an unintended lesson, too, as punishment often does: that he could be a real jerk sometimes.

PUNISHMENT CAN WORK, BUT . . .

I have seen two excellent, well-timed, to-the-point uses of punishment with animals, one while Cathryn Hilker, a gregarious big cat trainer and I were in a large enclosure with two young cheetahs at the Cincinnati Zoo. We were sitting on the ground petting the goofier of the two, a mellow male who stretched and purred while we stroked his spotted side. His brother suddenly swatted the trainer on her shoulder. *Bang*, she slapped his paw nearly at the moment it hit her, shouted "No," and jumped to her feet. Off the cat went. Up I stood, being as fast as neither a cheetah nor a cheetah trainer, even one in her seventies. The cheetah's swipe looked playful, but trainers never want a big cat to see them as a plaything, because once you're a plaything it's only a few steps to being a preything. The punishment worked because it was immediate, it was measured, and it stopped the second the cat did what the trainer wanted—to get its paw off her shoulder.

The second example was at the training school, where Walter, a young male buffalo, began to walk the student trainers around the compound. It was supposed to work the other way, but given his eight-hundred-pound bulk, Walter easily got his way. On one such promenade, the buffalo dragged the students, even though these fit young women dug in their heels and flexed their biceps, off a blacktop road and onto the lawn. He lowered his great black head for a grassy snack as two students tugged at his lead. "You better get him out of there, because he's just reinforcing himself for being a pain in the ass," a teacher, a former dolphin trainer, warned. The students, huffing and puffing, pulled some more, to no avail.

Walter happily chewed on. *Whack!* The teacher slapped Walter on the neck with her broad palm and grabbed his lead. The water buffalo, looking surprised, followed her back onto the road.

The slap, which was meant to startle, not hurt, the buffalo, did the trick. As with the cheetah, the punishment was immediate, fit the crime, and stopped the moment Walter quit eating grass and returned to the road. Still, the teacher warned the students to use the smack judiciously. First, because Walter would become desensitized to it, which would force the trainers to hit the buffalo harder to get his attention. Second, because whenever you use discipline, no matter how judicious, you draw down the trust account. And third, because punishment may provoke nasty side effects: apathy, fear, and aggression. None of those are conducive to learning. A scared or anxious animal doesn't make a good student. An apathetic dolphin at the bottom of the tank can't be taught a thing. A raging buffalo is in no mood for instruction. Trainers don't want any of punishment's side effects, especially the horn-in-human-flesh kind, but whenever they use it, they crack open this behavioral Pandora's box. Even revenge might fly out.

This is why Gary Priest, a former killer whale trainer who now oversees training at the San Diego Zoo and Wild Animal Park, warns against using even the slightest punishment. To drive his point home, he once scared the wits out of a conference room full of zookeepers and trainers with a video of an elephant attack on a young keeper. This keeper had earlier in the day smacked the female with his elephant hook, called an ankus. Later, as he busily shoveled poop in the elephants' yard, he set his ankus on the ground. In the video, you can hear an odd *tink-tink* sound off camera, which turns out to be the sound of the aggrieved elephant trying to snap the ankus in two. That was just the Hitchcockian foreshadowing of what was to come. Having no luck with the ankus, the elephant decided to break the keeper. That proved much easier as she grabbed him with her trunk and tossed him here, then pushed him there. The young man's screams brought other keepers to the rescue. "Get your animals to like you," Priest counseled. "Don't punish them."

That's an extreme example, but punishment produces the same behavioral problems in humans, to lesser and greater degrees, plus another: hatred, a potent cocktail of fear, aggression, and apathy. How you mix it is up to you.

I, like many members of my species, have been known to respond to punishment with small acts of revenge. When I waited tables, should a customer treat me rudely, say, snap his fingers at me or bark some directive, I'd smile and say I was sorry, behaviors the customer wanted. Then I would go to the dish room and scan the mugs. At this restaurant we served clam chowder in the same mugs we used to serve coffee. Little chunks of dried chowder often stuck to the bottom after they'd gone through the dishwasher, and so we waiters examined the mugs before filling them up to make sure they were chunk-free. Whiny, pushy, or drunk customers got the reverse treatment from me. I would round up mugs with globs and pour their coffee straight in. I'd be sure to keep their cups topped off, appearing extra attentive, so that they would not see what lurked at the bottom. On occasion, the hot coffee would loosen a calcified chowder chunk and it would bob to the top. I might get stiffed, but it was worth every forgone dollar, especially when the customers were businessmen who shrieked like little girls. Don't use punishment with your waitress. As Skinner and I proved, it could provoke unwanted behaviors, not to mention dishware.

———

Disregarding its side effects, punishment can, obviously, be effective and even produce results faster than using positive reinforcement. When I was in high school, one of the cool girls invited me to skip Spanish class with her. We slunk around the school some but mostly hid in the bathroom. Later that day, when we passed our teacher, Señor Scoby, in the hall, he noticed us. His eyes actually looked hurt through his enormous glasses. In short order the P.A. system crackled on and my name echoed through the school. It was the first time I had ever been called to the front office for misbehavior. There I was matter-of-factly informed by a pudgy, balding guidance counselor with a syrupy southern accent that I would be suspended for a day. Then he called my mother while I sat there, and informed her of my travesty. When I got home, my mother, who rarely yelled, let me have it. Like a smack on the water buffalo's neck or the cheetah's paw, my punishment was to the point, immediate, and startling. It also fit me to a T. I liked school, so it hurt not to go. I was a good student and class president, so I also cared about my reputation. The cool girl was cool in large part because she did not. I doubt the suspension had the same effect on her as it did on me. To tell you the truth, skipping hadn't been all that much fun, and the cool girl even turned out to be a bit of a bore, as people who don't care much about anything often are. I did the math. I went on to do some other sneaky things typical of high school girls, like breaking my curfew and acquiring a fake ID so I could get into bars when I was underage, but I never, ever skipped a class again.

What typically keeps punishment from working that well is our love of it. It's the first thing we humans think of. We discipline too reflexively, too absentmindedly, too lazily, and, typically, too much. We're not always clear what we are punishing someone for. Our timing is often terrible. We chastise long after an offense has been committed, sometimes when the offender is doing exactly what we'd like, when we should be rewarding them. We jump on one small behavior for a long list of grievances that might stretch over years. We batter and berate in the heat of the moment or merely because we are in a bad mood. Grouchy or emotional, we don't match the penalty to the crime, so the trespasser ends up feeling not contrite but wronged. That can make the punisher angry, and thus the punishment grows exponentially. We may not be interested in curbing behavior at all. We may find fault to punish out of revenge, or because we dislike someone, or to gain the upper hand.

We, like other members of the animal kingdom, punish to settle who's the boss. We higher primates, perhaps the most social of social creatures, are big on hierarchy. We want to know who's in charge. The answer is ever shifting, but the question is there in many of our relationships, fleeting and permanent, at work, school, and home.

Dominance is in the details. This is why spouses, parents, siblings, and bosses bother themselves with the littlest things. A friend of mine's husband blew up at her early in their marriage because, while making cookie dough, she creamed the butter with the flour instead of the sugar. The sugar is the way to go, but she hadn't ruined the recipe, nor the world. They had another knock-down drag-out over how she peeled carrots. What were they really arguing over? Who was in charge.

I, like my friends and lots of other spouses, unwittingly skirmished to win control of my own marriage by thrusting "my" way on Scott, such as taking what I thought was the quickest route to our favorite beach. I thought I was helping my husband by showing him how to chop onions, but what I was really doing was planting my flag. No wonder he so often resisted what I thought was well-intended advice, and why, when he resisted, I snarled. I would even, I realized, overlook something nice Scott did because it wasn't what I thought he should do. He once painted the inside of our coat closet a cheerful yellow. He rearranged the shelves. He hung up nifty pegs. This really pissed me off. I thought he should have cleaned the basement first. Worse, I told him so.

KEEP YOUR ANIMALS HAPPY

A traditional training technique is to withhold food from an animal, the thinking being that if an animal is famished, it will be more motivated to do as asked. This is a common way to work with birds, especially birds of prey trained to hunt. Trainers keep the hawks', falcons', and eagles' weight as low as possible so the birds are sure to be hungry and return to the hunter for a meal, not fly off into the great blue horizon, tempting as that must be. This is tricky in practice, though, because the difference between a hungry bird and a dead bird is just a few ounces—why hunters are forever weighing them.

Food deprivation is not a progressive method. New-school trainers don't withhold food from their animals, because it can cause far more problems than it might ever solve. The animals get their base diet every day no matter how the training sessions go. Progressive trainers also don't withhold toys, play sessions, or whatever makes the animal's life a good one. They want their charges to be healthy and happy. That's the best state of mind for learning. A ravenous, lonely, or bored animal is likely to be too anxious, cranky, and, consequently, distracted to concentrate on a lesson. Also, a hungry animal is more likely to be aggressive and bite. Anyone who's had to take a test on an empty stomach should understand why.

The human equivalent of food deprivation would be to cut back someone's daily required diet of affection, fun, approval, whatever they generally need to be happy. Examples would be parents who withhold all privileges until a child brings home a sterling report card, a spouse who gives the other the cold shoulder until every single chore on a long list is checked off, a boss who expects his employees to run themselves ragged for the least little pleasure, say, a fifteen-minute lunch break. These are ways to motivate someone with punishment, in this case by generally making them miserable to some degree. Any progressive trainer can explain why this approach is not effective, and that if you use it, your human animals will grow anxious, cranky, even bitey. Like underfed birds, they may, if you're not careful, simply waste away.

At the school, trainer Gary Wilson warns students always to guard against that deep instinct to boss another living being around, what he calls our "inner primate." Because we respond so to hierarchy, we assume animals will too. Some may; many don't. You only end up scaring or provoking them. This incorrect assumption is unfortunately encouraged by our best friend in the animal kingdom—dogs, a species which are generally very forgiving of our pushy ways—and by traditional dog trainers, including the megapopular Dog Whisperer, Cesar Milan, and those men in black, the Monks of New Skete. The whisperer and the monks advise that you establish yourself as the alpha dog. To do so, the monks recommend rolling your pooch on its back and growling over it, something I admit I did to Dixie as a puppy in our neighborhood park. That made me look like a nut job and taught my dog never to let anyone roll her on her back or her side, including a vet. When she ran into health problems, that became a problem.

The equivalent of an alpha roll with a killer whale would get a trainer killed. There is simply no pulling rank on a killer whale. Even if a trainer could, they wouldn't. The progressive trainers I met rarely use dominance. They don't need to. They want a cooperative relationship with animals, and the principles of reinforcement largely work regardless of who's alpha and who's beta. Mara Rodriguez, who works with the cougars at the training school, walks tall and squares her shoulders not so much to look like the boss as not to resemble prey. She suffers no guff, because if she did, the cats might start to think of her as dinner; but when she wants a cougar to do something, when she wants to *teach* a cougar to do something, positive reinforcement is her way.

We fear that if we spare the rod, all of humanity will go to pot. It may be our DNA, but certainly by the fifth grade most of us have learned that punishment makes the higher primate world go around. We are so convinced that discipline is the answer, that when it obviously isn't working, our instinct is, oddly, to lay it on thicker, yell louder, ground the kid for longer, don't talk to the husband for a week or two or three, dock the employee's pay.

Trainers who work with positive reinforcement have to overcome this human urge. Ken Ramirez at the Shedd Aquarium won't even let his trainers use the word "no" with the animals, ever. Given even the least little chance to reprimand, his trainers, he says, will eventually overdo it. They are, big brains be damned, *Homo sapiens,* after all.

Our humanness, our blind faith in punishment, is a central reason why Skinner did not achieve his ultimate hope, that operant conditioning would improve the life of his own species. He did, however, revolutionize the world of animal training as pioneering marine mammal trainers took his principles and went to work with buckets of fish. Karen Pryor was one of those early dolphin trainers. When she draped a whistle around her neck in 1963, she had worked only with dogs and ponies, using traditional methods. Training the dolphins with positive reinforcement, she found she could accomplish so much more. The method opened an exhilarating cross-species channel of communication. Moreover, this style made for a more enjoyable way of working, for the animal and for her.

In addition to the dolphins, Pryor began using rewards

with birds, red-footed boobies, at the park, then with her pets at home. Then something magical happened to her. Dolphin training began to creep into her everyday life. "I stopped yelling at my kids," she writes in *Don't Shoot the Dog!* "because I was noticing that yelling didn't work. Watching for behavior I liked, and reinforcing it when it occurred, worked a lot better and kept the peace too."

LOOK THE OTHER WAY

At my desk, I hear Scott on the stairs and then in the bathroom, blowing his nose, then his steps bounce back down the stairs. I open the door to my office, my fingers poised to pinch my nostrils closed if need be. The stink has vanished. Only a slight mustiness laps at my nose. I step in for a look. The bike clothes have vanished. He must have collected them, maybe taken them all the way to the washer! I notice something. The rug is a darker red where the clothes were. The sweat has leached into the cotton weave.

I call "Thank you" downstairs and manage not to say one word, though I'd like to, about the stain, not even a wisecrack. I'm rewarding the behavior I like (clothes taken away) and ignoring what I don't (that he left his garb there in the first place, that it discolored the rug).

That is what trainers do when animals make mistakes or misbehave or do anything they don't want to reinforce. They ignore it. There are some behaviors they can't disregard, such as a raging elephant, but if nobody or nothing is about to get hurt, a progressive trainer will typically look the other way.

This shocks many people because it is so counterintuitive

to how we normally deal with each other. We human animals are prone to the reverse, heaping attention on bad behavior and ignoring the good. Parents don't notice children who are riding in the car quietly, but let it rip when the little ones pipe up. Spouses don't think to say thank you for small tasks, but should someone forget to take out the trash, bloody murder may ensue. Employers take hard work for granted, but write employees up for the smallest infractions. Animal trainers turn that thinking on its head.

The shock for humans is that ignoring unwanted behavior does not create anarchy. The killer whales do not riot. The elephants do not start an insurrection. The polar bears do not revolt. The zoo does not go to hell in a handbasket, but rather the opposite. In fact, without one slap on the wrist, the bear presses his side to his cage for an exam, an elephant daintily dabs paint on a canvas, and a killer whale gently nudges a human onto his snout.

The trick is that ignoring unwanted behavior is only half the equation. The other half is noticing and rewarding what you want. The two go hand in hand. I trained my eye to notice when someone was doing what I *wanted*, whether it be airline ticket agents checking me in quickly or a friend asking me to lunch. It was beside the point whether someone was doing their job or their duty. Either they were doing something I wanted or not. If yes, then I smiled, made eye contact, and was sure to ask the friend out to lunch in return. I also realized some people were doing things I wanted that I thought I didn't want. I don't like to talk on the phone when I'm writing, so I rarely answered it during the day. Then I realized I was discouraging my friends and family from more than just telephoning me during the workday. I was discouraging them from staying in touch with me. I want them to stay in touch, and so I began to interrupt my writing and answer the phone occasionally, or at least make more of a point to call them back as soon as I could.

At home, I realized my husband was doing many things I wanted him to—washing the dinner dishes, keeping the car running, bringing in the mail—that I was taking for granted. I started thanking him. The same if he took out the trash. If he eased off the gas pedal, again, I thanked him. Likewise if he threw one dirty T-shirt into the hamper, even if our bedroom chair was buried under half his wardrobe. The point was to reward what I liked and ignore what I didn't. I kept mum about his five o'clock shadow. I would step over unpacked suitcases on the floor without one sharp word, though I did sometimes kick them out of sight under the bed.

I slipped up here and there, but my complaints and criticisms plummeted dramatically. At first it was hard to hold my tongue, but I got results pretty quickly, and good results reinforced my own efforts. As Scott basked in my growing appreciation, he shaved more and tailgated a little less. Moreover, the more positive I was with my husband, or, more importantly, the less critical I was, the faster his husbandly defensiveness faded away. When I asked Scott to do something, he was more responsive. His spousal deafness miraculously improved. He seemed more at ease. Maybe, in a way he hadn't before, he began to trust me.

SIX

THE HOW-TOs OF POSITIVE REINFORCEMENT

Exotic animal trainers don't toss monkey chow about willy-nilly. It's not Christmas morning down at the sea lion's tank every time the trainers get her out. The camel doesn't get an apple just for being him.

Trainers use rewards for a reason. If they don't, positive reinforcement loses its pop, or worse, can work against them by teaching a behavior they never intended to. A carrot in a trainer's hand can be just a pointy orange vegetable, or it can be a tasty morsel offered in friendship, or, if cannily used, the means to teach a rhino to back up on command.

That is why I did not morph into Ms. All-Around Nicey Nice, showering anyone and everyone with smiles, compliments, gifts, and gratitude. Nor did I become a model 1950s wife, greeting Scott every evening with a kiss, a martini, and a home-cooked meal. However, I became more pleasant, to some extent, just by ignoring behavior I didn't like, especially my husband's. No nagging (well, almost none) equals nicer.

With Scott, I was still affectionate and attentive in general, but, like a trainer, I reserved positive reinforcement for when he *did* something specific I liked, that I wanted to encourage: walking the dogs in subzero temperatures (isn't that why God gave us Minnesotans?); getting rid of a great heap of branches, grass cuttings, and leaves from the garden (where, I don't ask); and greeting *me* with a martini (gin, up, dry, three olives, chilled glass).

Shamu taught me to be specific with rewards, which sounds easy enough but so often is not. The principles of positive reinforcement are simple, but as any animal trainer will tell you, applying them is slippery business. I've seen student trainers freeze, chicken neck in hand, as they did the math of a behavioral equation in their head, one that had seemed so obvious until they actually stood before a lion. The students had to learn the finer points of using positive reinforcement, and so did I.

TIMING

During a visit to my family in Cincinnati, my two nieces ask me what Scott will do while I am gone. "Wear underwear on his head," I say. "Sleep in a big pile of dirty clothes. Lick the inside of pizza boxes." This is a running joke for which I obviously exaggerate, but the man has given me some material to work with over the years. I have come home from trips to find towels ripening in the bathroom, pans caked with omelet makings in the sink, and unopened mail piled here and there.

After this trip, though, I arrive home, open the back door, and step into a pristine kitchen. The stainless counters shimmer. The coffeemaker has been wiped clean. There is nary a greasy pan in our big porcelain sink. "Oh my god," I blurt as I circle the island. "Thank you, thank you," I call to my husband, who's meandered off. I've never come home to such a clean house. This demands positive reinforcement—*now*.

———

Animal training is in the timing. You can have all your hand cues down and know the natural history of your species backward and forward, but if you are even a hair slow, you're going to have a problem teaching even the most willing animal to do a trick.

Good trainers let animals know the exact moment they get a behavior right. Not a second before and not a second after. When a trainer teaches an elephant to trumpet on command, the second it hits a note, *bang*, the trainer lets the elephant know "That's it, that's what I want." This is why trainers use whistles and clickers (those plastic noisemakers also called crickets) or chirp "Good boy." These are all what are known as markers or bridges, basically a quick, clear way to inform the animal, "Bingo, treats incoming." Trainers use bridges as a way to instantly communicate to an animal when it has done the desired behavior. Without a bridge, a trainer is left to fumble with a reward, and by the time he's tossed the fish or proffered a banana chunk or kibble, the animal will be doing something else. That something else, a sea lion's swimming to the side of the pool or a dog's jumping up from a down, ends up being what the trainer reinforces. So without a bridge, a trainer's timing is always slow. With it, her timing can be exact. That is why when pioneering marine mammal trainers began using whistles it was a major breakthrough.

SUPERSTITIOUS BEHAVIOR

When trainers accidentally teach an animal something, they call it a superstitious behavior. As animals are such quick studies and notice absolutely everything, it's not hard to do. This can happen a number of ways. When teaching a tiger to sit, if a trainer's timing is a wee bit slow, he might reinforce the tiger as it sits and growls. All the trainer wanted was a sit, but now the tiger thinks it's a package deal.

Or, by training the sit in the same spot over and over, the trainer can accidentally teach the tiger that a sit is done only in the southeast corner of the cage, nowhere else. That's why trainers practice new behaviors at various locations, even within the same cage, and even with the animal facing different directions.

Or a trainer can teach a superstitious behavior by adding a cue she didn't mean to. If she unwittingly tilts her head while signaling with her hand, the animal will come to expect both. So next time the trainer gives the hand cue but doesn't tilt her head, the tiger thinks, "Huh?" while the trainer thinks, "Why doesn't this tiger sit?" Each looks at the other, deeply puzzled.

Humans likewise connect dots on the spot that they shouldn't, and likewise are prone to superstitious behavior. There's the obvious stuff, the lucky objects and garments, the tossing salt this way or that. Even the most rational of us have one or two. Kids have tons around sleeping. When I was little I had to have my "baby blanket," a crib-sized pink quilt, over my chest and tucked just under my chin. I was convinced this protected me from vampires and zombies and Barbie dolls that grew fangs in the dark. Why? Because every night I slept with my baby blanket, I never got bitten by a vampire or a zombie or a Barbie with fangs.

Then there are the little bits of behavior, habits, we accidentally train ourselves. I cannot sit down at my computer in the morning without a cup of coffee. I've taught myself I can't start typing without caffeine. Of course I can. When my mother cut back on cigarettes, she had to face a long list of activities that she thought were impossible without a lit butt in her hand. How would she, she wondered, talk on the phone without a cigarette? Or go grocery shopping? Or drink coffee?

Superstitious behaviors are basically accidentally reinforced behaviors. The term, to me, underscored how easy it is to teach something you didn't mean to, to yourself or someone else. If a behavior seems to produce a good result, whether it actually did or didn't, that behavior is going to stick. If the sky looks iffy, I take my umbrella, because if I don't, the heavens always seem to open and shower down upon my bare head. I know my umbrella has nothing to do with whether it will rain or not, but somewhere in the recesses of my brain, my past experience has linked the two (umbrella in purse = no rain, umbrella not in purse = monsoon).

That superstitious behavior is harmless, though my purse grows heavier. Some superstitious behaviors, though, can get in people's way. I had a student who spent his semester falling further and further behind on his assignments. After each class, he'd approach me to discuss his increasing lateness and the stalled status of his assignments. I grew to dread these predictable, aimless talks. Then I realized I was reinforcing the behavior, and I suspected it had become a superstitious one for him, that to complete work in a class, he had to get behind and then talk to the teacher about getting behind, and only then, under great pressure and duress, could he do his assignments. It was his equivalent of having a rabbit's foot in his pocket. I checked with another of his professors, and sure enough, the student was behaving the same way in his class. This superstitious behavior was not harmless, and his grade, at least in my class, showed that.

There is a beautiful simplicity to the animal mind. Nature has designed animals to live in the moment. It's not that they can't think ahead or behind, but animals immediately link events that happen simultaneously. If a tiger puts his front paws on his cage and gets an instantaneous click, followed by a chicken neck, the tiger's brain makes a note: Paws on the cage = click = chicken neck. If just as a dolphin threads his nose through a small hoop he hears a whistle, then gets a mackerel, he likewise jots a mental note: Nose through toy = whistle = mackerel.

Whenever a trainer has trouble, the first culprit to consider is timing, usually that they are a few beats behind in tooting the whistle or clicking the clicker. That's often true because humans are the slowpokes of the animal kingdom. Most animals are fast, scary fast. I experienced countless examples of their uncanny speed. As instructed by one gregarious trainer, I ducked into a bobcat's roomy cage and sat down on a dusty truck tire. The bobcat, in a smaller cage off its main enclosure, watched me. I didn't know what the trainer had in mind until, with a smile, he opened the gate between me and the bobcat. One split second the cat was a good twenty feet from me, ears pricked. The next it was perched softly on my shoulders, rubbing against the back of my head as if it were a housecat. I didn't have time to yelp. I didn't even see it leap. I must have at least bugged my eyes as the wild kitty nuzzled me, because the trainer, outside the cage, laughed his head off. If I were a deer I'd be a dead deer.

No wonder the students often squeezed their clicker a few beats too late, thus marking the wrong behavior, teaching a sit when they meant to teach a circle. Savuti the hyena at the training school was such a font of behaviors that if he didn't hear the click immediately, say, for picking up a log in his freakishly strong jaws, he'd try something else, hold up a paw or look over his shoulder and smile. Usually by then, students new to working with Savuti had finally managed to squeeze their clicker, about four behaviors after they meant to.

With people, I could not be as precise as a trainer, but I began at least to think about my timing. Ideally, I'd reward someone the moment they did something I liked. That's why applause is such a thrill; it's not only a response, but an immediate one. This is the reason writing is not such a thrill. The reward comes months, even years later, if then, but certainly never when you are actually writing. By the time an article or book has been published, I'm typically gnashing my teeth over a new project and the former is a distant memory. That's like giving a dolphin a big tuna about two years after a flip.

Humans, of course, don't need the immediacy animals do. We are accustomed to getting a payoff down the road; still, the sooner the better, especially when learning something new. Sooner is tough to do because human animals do things you like when you're not around. You can't help rewarding them after the fact. My policy became to reward behavior I wanted the very first chance I got. To do this, I had to give up some of my natural dithering. I RSVP to dinner invitations ASAP. If someone e-mails me a compliment about my work, I e-mail back a thank-you note right away instead of letting it languish in my in-box. If a present arrives in the mail, I open it up and then phone the giver. And when given the rare chance to reinforce someone in the moment, I jump on it.

Thus my demonstrative behavior just now. I leave the kitchen in search of Scott to give him a big kiss. I find him in the foyer, where I eye a stack of unopened mail. I look right past it and pucker up.

WHEN NOT TO USE IT

One February morning at SeaWorld, I watched as a male beluga whale resembling a mini Moby Dick attempted to hoist himself onto a poolside scale. It seemed an impossible task, heaving that alabaster bulk out of the water. Belugas aren't near the jumpers dolphins are. The whale rose out of the pool, tapped his chin on the end of the scale, then slowly slid back into the pool. His ghostly silhouette sank below the surface. The beluga had given it a shot, but the trainer's whistle remained silent. Not a single fish was tossed. To a trainer, trying does not count—only doing.

Why not? Because the trainer wanted the whale on the scale. That's, as a trainer would say, the criterion. If the trainer had tooted and pitched the beluga a herring in the pool, he would reinforce the whale's *trying* to get on the scale. It doesn't matter if the whale tried in earnest or made a lazy, half-hearted attempt. Either way it's still just trying.

When the trainer first taught the whale this behavior, he might have started by getting the beluga to touch the scale with his chin, for which the whale would have been rewarded. But this beluga already knew that behavior, which quickly became apparent. After a brief pause, the trainer cued the whale a second time. He once again rose from the tank, water rushing from his milky-white sides, heaved onto the scale, and kicked his tail up jauntily as if to say, "Take that." There he stayed while the number on the scale climbed. When it read a steady two thousand pounds, the trainer tooted his whistle. Back in the pool, the beluga opened wide for his paycheck— one neatly tossed fish.

How often was I reinforcing *trying* when what I wanted was *doing*? If Scott opened the refrigerator door and tried to find a jar of salsa, I'd come to the rescue. If my mother told me she was trying to quit smoking, I praised her. If a friend or relative mentioned changing careers, I'd root them on. Animal training showed me, once again, how my good intentions might undermine me, and why cheering on friends and family didn't always get results. If you reinforce trying, that may be all you get.

Trying in humans, I found, often translates into talking about doing such and such, sometimes ad infinitum. People get a lot of reinforcement for verbally trying. I'd just have to say to my friends I was thinking of studying Spanish, and they'd say, "What a good idea! How smart of you." All that, and I didn't have to learn to conjugate one Spanish verb, just say I would. *Olé.*

Scott talks about joining group bike rides. My friend Dana talks about losing pregnancy weight going on seven years since the birth of her youngest. My mother, who has been quitting smoking since the Pleistocene Age, and I talk about how she might give up the very occasional cigarette she smokes. I suggest the patch. She'd rather take vitamins or try drinking more water. Again, I suggest the patch. We talk about it.

When Scott mentioned any group activity, I'd ooze encouragement. I'd tell my friend she looked great. I'd cheer my mother on. Then I realized that in these conversations, I was only reinforcing *talking* about group bike rides, shedding some belly flab, or tossing out the cigs. I was a trainer tossing a fish to a beluga that had only stuck his chin on the scale. That's fine unless I really want my husband, friend, and mother to do these things. In that case, I should save my encouragement. I really want my mother to lay off the smokes. I would like my solitary creature of a husband to do something social. However, Dana truly looks fab, pregnancy weight or no. I will continue to discuss cellulite with her and what kind of bathing suit she can get away with. Dana says a skirted one-piece. I say a bikini.

SIZE MATTERS

Before students at the training school would work with the big cats, they'd hack up chicken necks into pearly pink chunks. They would trim bananas, grapes, and apples into coin-sized bits before getting a capuchin out for a session. They'd slice sweet potatoes into wedges and carrots into discs before taking the water buffalo for a walk.

The trainer's rule of thumb for how much reinforcement to use is the smallest bit that will do the job. If a monkey will train for a piece of banana, don't offer the whole fruit. If the water buffalo will work for a cube of sweet potato, don't hand him the spud. This is for practical reasons. It takes a monkey longer to eat an entire banana than just a piece. Training will come to a screeching halt while you wait for the monkey to polish off his treat. Add to that, a monkey will fill up quickly if given whole bananas, and lose interest in training. You'll get maybe three or four behaviors out of her before she's through. With morsels, her stomach will fill up more slowly and she'll be a willing student for longer.

I borrowed the trainer's less-is-more gauge for how much positive reinforcement to use with humans. I found that this assured I'd never overdo it. No showering Scott with kisses for putting one sock in the hamper. If a matter-of-fact "thanks" did the trick, I'd go with that. Thinking like this kept the gratitude in line with the task. If I bear-hugged Scott for bringing in the mail, he would find that less reinforcing and more patronizing. This is one way humans differ from all other animals. An animal will never find an extra-big reward insulting, but a human might. Likewise, this measure kept the reinforcement in line for myself. I wouldn't be tempted to overdo it by way of sarcasm, the equivalent of handing a water buffalo a big, but rotten, piece of sweet potato.

Sometimes, though, a whole sweet potato *is* what is called

for. The bigger or more difficult the task, the bigger the prize required. Trainers have to make it worth the animal's while. In *Don't Shoot the Dog!* Pryor writes that at Sea Life Park in Hawaii the false killer whales (which resemble orcas but are slenderer and nearly all black) would not perform an epic twenty-two-foot high jump straight out of the water for the standard reward of two smelts. They demanded higher pay for such a feat—a large mackerel.

Again, this seems painfully obvious, but how often we human animals don't match the reward to the task. At a daily newspaper I worked for, I was drafted into filling in for my boss whenever he went on vacation. This was meant as a compliment, of a sort, but the upshot was that my workload would double, some of my coworkers gave me attitude, and I had to go to endless news meetings, many of which had more to do with male bonding (boys, do it on your own time) than with what stories would run on the front page. For all this, I got twenty or so extra dollars a week, I'd say about a half smelt to a false killer whale—a really puny, dried-up half smelt. The first second I could I wriggled out of filling in for my boss. I wasn't going to jump that high for a crummy little fish.

Another rule of thumb for trainers is that you have to give the animal something better than what it already has. If a killer whale is having fun playing with a ball or a poolmate, the reward you offer for its attention has to be worth more to the animal than goofing around. If a fennec fox loves to nap in its den box, and you want it to come out and train, you need a treat that appeals to the little canine with big ears more than a cozy snooze.

In these two cases, what each animal is doing is what is called self-reinforcing behavior. Put simply, a self-reinforcing behavior is gratifying in itself. They come in all sizes for animals and people alike, from sailing across the Atlantic to popping bubble-wrap bubbles. (Even my dog Penny Jane enjoys the latter. She squeezes the sheet between her paws and bursts the bubbles with her teeth.) If you enjoy the mere act of doing something, whether shopping at Chanel or clearing your throat, it is self-reinforcing. The more self-reinforcing a behavior, the harder it is to stop or distract someone or something from it. You'll need a reward that is more reinforcing than the behavior. That goes for the entire animal kingdom.

At a restaurant I worked at, the one with the chowder/coffee mugs, I would aim a whipped cream canister at a piece of pumpkin pie, press the nozzle, and *splat*, a pool of runny cream would douse the dessert. The whip in the whipped cream had disappeared once again into the cooks, who had inhaled all the nitrous oxide out of the canister. They would suck on the canisters while making salads or tossing chicken breasts on the grill. At least then they were on the line. Plenty of times, I'd turn in an order to an empty kitchen. They would all have absconded behind the restaurant with a canister or two to get high. Talk about self-reinforcing.

Pleading, berating, and nagging, obviously, got us waitresses nowhere. Getting stoned on the gas was so much fun it was worth our punishing screeches. To get them to lay off the canisters or keep them in the kitchen, we needed a really big mackerel—free drinks from the bar. On a busy night, we'd sneak a steady stream of beers and cocktails to the pass window. Call us enablers, but we got the behavior we wanted—

our orders cooked and on time—and maybe saved the cooks some brain cells, to the detriment of some liver cells. There's something about waitressing that brings out the instinctive animal trainer in you.

JACKPOT!

Scott dashes upstairs to pack a bag. He just got word that our tenants in Boston, about a two-hour drive away, have moved out but left the heat on and a window open. It's late November. Our condo, according to our neighbor's report, is heating the entire block. Scott has dropped everything on the first day of his Thanksgiving break to race south and rescue our gas bill. As he blasts through the kitchen on the way to the basement to find a suitcase, he grumbles, but he's largely being a good sport. Time, I realize, for a jackpot.

A jackpot is what it sounds like. A big, fat, juicy serving of positive reinforcement. Ideally, it's a surprise, just like when a slot machine releases all those delicious coins. The bonus, the grand prize, those are jackpots too.

Trainers give jackpots when an animal makes a big breakthrough. A jackpot is like a stadium full of cheering fans all yelling "Hooray!" Some trainers use it as a wake-up call, too, if an animal isn't responding. Karen Pryor once threw a sulking dolphin two fish for free, which startled the animal out of its apathy. Ideally, the jackpot coincides with the action you are encouraging, but, as always, that's easier to do with animals than humans.

With the heating emergency, I find timing on my side for once, because my point is not to reinforce Scott's going to Boston but that he's being a good sport about having to, which is what he is doing right now.

This is clearly the moment to throw a whole bucket of fish into the pool. I run upstairs and pull a Christmas present for Scott out of my closet, a stereo bauble. I dash back down to the kitchen and set the gleaming white gizmo on the island. I hear him trudge back up the basement stairs. The door creaks open, he steps into the kitchen, his eyes fall on the shiny new piece of gear. He freezes. His jaw actually drops. He loves his jackpot so much, he delays leaving for an hour to play with it. The man who doesn't know how to change the ringtones on his cell phone has this thing up and playing in seconds. Down in Boston, hot air gushes out of our open window, but I don't hurry Scott. I figure part of the jackpot is enjoying the jackpot.

MIX IT UP

I stand hip-deep in a pool of cool blue water. Each time a small wave laps me I shiver despite the bright sun, despite my wet suit. Before me, a dolphin's gray rostrum pokes through the pool's light chop. This is the reason the water temperature is fifty-five degrees and salty. As a trainer next to me instructs, I point with one finger to my left. I have no idea what I've just asked the dolphin to do, but she does. The animal vanishes underwater. I scan the pool for the jut of her dorsal fin or a quick-moving shadow. Nothing. Suddenly she explodes from the middle of the pool like a geyser.

As the dolphin shoots heavenward, the trainer sounds his whistle. She splashes back into the pool and surfaces right in front of me. The trainer passes me a handful of ice. I look at the ice and then at him. "Give it to her," he instructs. I turn to toss the chunks into the dolphin's mouth, then hesitate for a moment when I notice her smile is lined with small but obviously sharp teeth. There are lots more of these teeth than I ever imagined. I risk a finger or two and pitch the ice between her jaws. As the dolphin gargles them, she makes a sound like a daiquiri being blended.

Ice is just one of the many rewards trainers give the dolphins. They also give them Jell-O cubes, rubdowns, and toys. They treat the killer whales with a squirt from a hose or a look in the mirror. They toss the polar bears frozen chunks of watermelon. At the training school, Mara Rodriguez uses what she calls her "happy voice" to reward the cougars. She also gives them a rub on the head or moves closer to the cats, which they like.

The more reinforcers the better, because even prizes, if predictable, can lose their shine. A variety ensures that an animal won't get tired of any one kind of reward. It also keeps training, as well as the animals' lives, from becoming routine. Nonfood reinforcers, such as a rub on the chin, add to the variety but also have the advantage of extending a training session past when an animal's stomach is full.

This is where knowing your animal comes in handy, because reinforcers, like punishments, are subjective. One organism's reward is not another's. Birds of prey, which self-preen, are not inclined to find a scratch from a trainer much of a perk. Macaws, which preen each other, typically do. Yet a parrot raised in the wild may not cotton to a scratch from a trainer as readily as a hand-raised parrot. A reward's value can change depending on an animal's age as well. A young animal might prefer play as a prize, then lean more to treats as it ages. Reinforcers can also wax and wane depending on circumstances. A happy animal may like a pat, a worried one not.

I found it a good idea to mix up reinforcers for humans as well, if only because it kept me from having to say "Thank you" constantly. I thought of a list for Scott: smiles, hugs, kisses, compliments, head rubs, and presents (especially stereo and bike gear). Relieving him of one of his regular tasks, say, taking out the garbage, qualifies too. Dana loves magazines, so if she makes time to meet me at the beach, I sometimes bring along one or two for her to page through. I give Hannah an occasional bauble in exchange for reading my raw copy. If my mother calls, I reward her with funny stories about me, my husband, or my dogs. My dogs themselves, especially Miss Dixie, are bonuses for a few of my friends. On the rare occasion I don't bring my Aussie with me to meet my friend Ray for lunch in the park, he will look at me as disappointed as a sea lion that flipped and didn't get a squid, and ask forlornly, "Where's Dixie?"

Thinking back, all my good bosses have naturally used a range of perks, which are especially effective because the predictability of a paycheck can undercut its power. It starts to feel like a free feed, which is nice, as any camel would tell you, but not always the most powerful motivator. Two of my bosses made a point of giving me story assignments they knew I would like. Another, an executive editor, left Post-its on our computers with compliments scribed in his tiny handwriting when he especially liked a story, what we reporters came to call "Lou notes," as in "You got a Lou note, you bastard." Another boss, Bob, would occasionally send us home early on a Friday afternoon. "Get out of here," he'd suddenly announce, at which we would gleefully grab our purses and skedaddle. We loved Bob.

VARIABLE SCHEDULES

When animal trainers teach a new behavior, they are generous with the treats, tossing one each time the animal makes the grade. But once an animal has the new move down, trainers may taper off the rewards and use a variable schedule of reinforcement. That is, sometimes the animal gets a reward for the behavior and sometimes not. This isn't done to economize on squid or monkey chow. This a powerful way to maintain behaviors.

As Skinner proved, trainers demonstrate, and the whole animal kingdom makes clear, an organism will repeat actions that produce good results. Reinforcement makes the world go around. Living things are so inclined this way that the good thing doesn't have to happen every time, or even close to every time. Just often enough to keep the animal motivated. Sometimes chimps find fruit on the jungle floor and sometimes they don't, but it's always worth a look. Sometimes lions bring down a wildebeest and sometimes not, but it's worth a run. Sometimes I pick a winning horse at the track and sometimes—actually, most of the time—I don't, but it's worth plunking down two dollars to me.

As anyone who has gambled, from the bingo table to the fifty-dollar window at the track, knows, a schedule of variable reinforcement is mighty powerful. Just a very occasional win keeps you betting. In all the times I've gone to the track, I've had maybe one or two days when I came out ahead. Still I study my racing form, examine the horses and the jockeys in the paddock, make my bet, then step up to the rail and hold my breath. Any race might deliver a pellet, potentially a really big one, like the time I won back-to-back races at Santa Anita for a total of forty-eight dollars.

Variable reinforcement is a double-edged sword, in that it maintains behavior you don't want as well as behavior you do. People unwittingly use variable schedules of reinforcement to teach dogs to beg (giving in to that pathetic stare every once in a while), kids to pitch fits (giving in to the screaming every once in while), and spouses to nag (giving in to the badgering every once in a while). That is why I kept nagging my husband, because every once in a great while it worked, and Scott would unpack a suitcase or shave.

You can also make the behavior you don't like so much more intense. If you occasionally hand over a snack only after the dog has begged for nearly the entire meal, you've taught him to hang in for the long haul. If you occasionally fold only at the height of a long tantrum, you've taught the kid that duration and earth-shattering volume do the job. If you occasionally take out the trash after a full week of carping, you've taught the nagger that seven days of pestering is the charm.

I did not consciously adopt a variable reinforcement schedule in my relationships, but I realized I didn't have to reward Scott or anyone every single time they did something I liked. In fact, in some cases I shouldn't. The thanks and appreciation might become meaningless for them, not to mention tiresome for me. The trick was to make sure I rewarded what I liked often enough to keep people doing it, to maintain those behaviors.

Moreover, the power of a variable schedule explained why people did the things they did. I finally understood why my mother never quits smoking. By smoking a few cigarettes every so often, she keeps the habit alive, if not more so, than if she burned through a pack a day. I understood why a friend stayed in such a dismal relationship. Her boyfriend would toss her a mackerel just often enough to keep her waiting around for another.

I finally understood why I couldn't quit gardening despite much heartbreak and cussing in my yard each summer. Though the coneflowers always look as anemic as runway models, the red fountain grass refuses to bloom, and at least one rhododendron croaks with the drama of a silent screen star each season, just enough plants thrive to keep me digging, planting, and fertilizing though I swear each August I'll stop. Damn the variable reinforcement of those showy day lilies and hydrangeas that explode with flowers occasionally! And those daffodils that nudge the raw earth out of the way some springs, egging me to pick up my shovel one more time. That garden has clearly trained me.

SEVEN

BABY STEPS

A trainer would never just hand a brush to an elephant and say, "Go to work, Monet." At Have Trunk Will Travel, a private elephant compound in Southern California, a trainer first teaches the elephant to curl its trunk around the brush. Next, a trainer instructs the elephant to dip the brush in a dish of water. Then to dip the brush in the water and in the paint. Then dip the brush in the water, then in the paint, then dab it on the canvas. And so on, until the elephant churns out one work of abstract expressionism after another. For some reason, a trainer told me, the elephants often have trouble learning to dip the brush back in the water before using a different color of paint. I think Jackson Pollock may have had that problem too.

These phases are what trainers call successive approximations, the baby steps toward learning a whole new behavior. Trainers basically construct a behavior from the bottom up, starting by teaching an action that will be the foundation and then building from there. For example, once the really tall teenage student trainer at the school slouched enough so that Rosie the baboon wasn't unnerved by his looming frame, he taught her to sit on a skateboard and remain calm. Next he pushed the board back and forth gently with his hand as Rosie perched on it. When she didn't mind that, he trained the baboon to stand on all fours on the board while he rocked it. Last, he taught Rosie to put one of her black-soled feet on the ground and push. And so another shredder hit the pavement of America, only this one has a bright pink bottom topped by a jauntily upright tail.

The idea of approximations is nothing new to humans. This is how we teach all kinds of skills, from how to read to how to play tennis. But watching exotic animal trainers, I realized we humans too often expect the whole enchilada on the spot from friends, family, employees, and coworkers when it comes to changing behavior. I knew I had. I thought that if I told Scott how much his jeans, T-shirts, and fleece pullovers piled on the footboard of our bed bothered me, that he would change overnight a habit that had been a lifetime in the making. If he didn't put away the whole teetering stack, no positive reinforcement for him. In fact, I gave him the opposite: harrumphing, loud sighs, wisecracks, and increasingly dramatic reiterations of why it irked me so. My expectations, not to mention my responses, were not only unreasonable but counterproductive. The proof was in the resulting behavior—snarling at me and not putting one single T-shirt away.

When an animal won't do a behavior, won't take the next step, then it is probably too big. Likewise, expecting someone to change overnight is entirely too big an approximation. Humans, unfortunately, put a lot of stock in the gigantic leap, the turnaround, the overnight success, the total makeover, especially when it comes to ourselves. Every year we write up a list of gargantuan approximations, also known as our New Year's resolutions, and we all know how well those work out. We just can't go from being person A to person B suddenly, as much as we'd like to believe we can. Why not is very obvious to an animal trainer.

The sink-or-swim approach is another example of using enormous approximations. It is too often our modus operandi, though this approach often sinks us. At many newspapers, it's customary to bury new reporters with a ton of assignments to get them up to speed quickly. It also takes a few years off your life, as I can attest. Nothing makes the blood pressure spike like working at warp speed and trying to impress a new boss when you don't even know where the delete key is on your computer. You may get up to speed more quickly than if you eased in, but at what cost? This approach is how a lot of employers burn out their new workers in their first few months on the job. They take an energetic, eager employee and in short order turn them into an exhausted, resentful underling.

Dauntingly—if not impossibly—enormous approximations turn up everywhere, sometimes in places you'd never expect. When I was five, my mom took me to swimming lessons at a pool so big the deep end was monitored by a lifeguard in a small rowboat. All I remember of the lesson is this: To accustom us to being underwater, the teacher told us to hold our breath, pushed us under, then put her foot on our heads and shoved our small bodies to the bottom. I still remember clawing at her foot with my little fingernails. Maybe her system worked, because I, in spite of her half drowning me, learned to swim and am not afraid of the deep end. However, she did teach me to be afraid of swimming teachers. I never took another swimming class.

The trial-by-fire approach is not only an unreasonable approximation, but an awfully lazy way to teach. It's not teaching, it's "Just do it because I say so." There's something in it that smacks of dominance to me. We wouldn't expect a sea lion to understand if, out of the blue, a trainer commanded the animal to salute. "Come on, you're a marine mammal. Figure it out." Spouses soften that approach to "Do it because I asked you to" or "because you love me." Bosses say do it because it's your job. Parents say do it because I say so. You can ask or order, but if you demand too much all at once, you're not likely to get what you want. Slobs cannot instantly become neatniks, procrastinators cannot overnight become time hounds, and speed demons cannot suddenly become cautious drivers no matter how much they respect, like, even love you.

———

At home, I stopped expecting instant radical changes from my husband. Rather, I began to praise small bits of progress. I applauded if Scott drove a few miles per hour slower or kept me waiting at a restaurant less time than normal. When he shifted his teetering pile of clothes from the footboard to the top of his trombone case, I considered that a noteworthy improvement. When the pile occasionally shrank, likewise.

Using approximations not only made my expectations more realistic but helped me analyze a behavior, to see the little parts that would lead to the sum, to understand the snags along the way. I do this with myself. In fact, I always had with writing, always setting my sights on penning single sentences or working out one problematic paragraph, avoiding thinking of the whole article or book. But I hadn't applied this logic to my personal life. Now, if I feel overwhelmed by a task, especially if I am procrastinating, I break it into steps. My physical therapist, my doctor, and all my friends told me to go to yoga. Every time I thought of it, "doing yoga" was an overwhelming life change. I never went to exercise classes. I owned no exercise clothes. I called one yoga teacher to ask about his class, and he told me that "I needed to be in my body." That was the one thing I had going for me, that I was already in my body. I didn't need help with that, so I didn't sign up for his class. However, my body hadn't had its muscles thoroughly stretched since gym class in 1975. I would need to extend my hamstrings if I was going to "do yoga." And so another day would pass without my moving any closer to the lotus position. Finally, I broke it down into successive approximations: get yoga clothes, get yoga mat, find a class, sign up for a class, go to a class, and, way down the line, assume the lotus position. I began with baby step number one, shopping for yoga clothes, which would be fun for the forager in me. That's as far as I've gotten, but it's one step further along than I was.

I realized my campaign to convince my mother to get hearing aids might do with some approximations. For the past few years, I've tried nagging her to do so, though this is as sure a way to make my mother mad as to call her a "hillbilly." "They're too expensive," she'll say. "I'll help pay for them," I say. "Everybody I know who has them hates them," she retorts. "They've made big advances in recent years," I'll say. At about this point, especially if I've handed her another newspaper clipping, she usually just glares at me. Once, out of desperation, I tried something along the lines of "Only hillbillies don't get hearing aids," which was like poking a camel in the mouth with a cattle prod. As you would expect, the camel charged. She snapped at me.

I kept rephrasing the suggestion, trying out different tones, but I mostly repeated myself. The last time I started in on my mother, midpitch it dawned on me maybe I was asking for too big an approximation. What would be the first step to getting hearing aids? Getting a hearing test. That's what I set my sights on. She hasn't taken the test yet, but at least the suggestion of one does not irk her as much as mentioning hearing aids does. That is progress. Maybe I need to break the behavior down even further. I might get her to just make an appointment for a hearing test. Then get her into the car to go to the appointment, and so on.

GO BACK TO KINDERGARTEN

Just as with positive reinforcement, there is a science to using baby steps, even rules. Karen Pryor, the scientist, writer, and pioneering dolphin trainer, lists ten in her book *Don't Shoot the Dog!* These are taught at the training school. Many professional trainers can tick them off, especially marine mammal trainers. In fact, they can probably say them in order.

All these rules apply to humans, but only five made my short list. Not using too big an approximation, as described above, is numero uno, on my and Pryor's lists. That is obviously a keeper for me. Number two is this: When a behavior deteriorates, go back to kindergarten.

One summer afternoon at the school, I watched as a student, a matter-of-fact young woman with pink streaks in her hair, tooted a whistle three times, the cue for Harrison the Harris hawk to fly from her gloved hand to another student just out of sight behind a set of bleachers. Harrison twitched his tail feathers and then lazily unfolded one wing as if he was toying with the idea of flying, then, thinking better of it, folded the wing in tight. The student tooted again. Harrison stayed put and stared straight ahead with his copper eyes.

Before going on vacation, the student had been teaching Harrison to circle the bleachers and back to her on stage. She'd nearly gotten the hawk to swoop around the entire bleachers, but now he wouldn't even leave her big leather glove. The student sighed. "I lost my training behavior. What do I do now?"

"Take him back to kindergarten," a teacher nearby called.

When behaviors fall apart, for whatever reason, trainers take a few or more steps back in the training process. Sometimes all the animal needs is a quick refresher course, sometimes a longer one. The point is to back up, not keep pushing on, making you and the animal increasingly frustrated. In Harrison's case, the student who was behind the bleachers walked closer to the stage where the hawk could see her. The student with pink streaks tooted her whistle again. Harrison threw open his wings and with a few flaps alighted on the other student's glove.

We human animals go back to kindergarten when we study for a test or brush up our French by listening to tapes in the car. I go back to kindergarten at the start of every ski season by taking my first few runs down forgiving slopes. In fact, I've never quite gotten out of kindergarten when it comes to skiing, or tennis, or knitting, for that matter. So it's not hard on my ego to go back.

Going back to kindergarten emphasizes how behavior is never absolutely fixed, that it naturally shifts with time and circumstances. Just because your dog comes when called now doesn't mean he will forever. You may have to reteach him to come bounding at his name. Just because your kid has learned good table manners doesn't mean he will henceforth always be a proper diner. You may have to reteach him to eat pancakes with a fork and not his hands. Just because you've taught your spouse to call when he'll be late for dinner doesn't mean he'll phone every single time. You may have to reteach him to give you a buzz, or, a step before that, to carry his cell phone with him. Just because you learned to sail as a kid doesn't mean you will always know how to bring the boat around. You, like me, may have to reteach yourself when to lower your head so you don't get brained by the boom.

Moreover, going back to kindergarten is a much-needed antidote to the human drive to butt things head-on. My poor husband expects himself to play tennis as well as he did the last time he picked up a racket, even if it was five years ago.

He swings away, getting more and more frustrated. If only he would go back to kindergarten. It might be hard on his ego, but so much better for his game, not to mention his mood.

ONE BABY STEP AT A TIME

Number three is to train one aspect of a behavior at a time. In other words, if you are using successive approximations, keep them successive. It's not enough just to break a behavior into small steps.

Pryor offers the example of teaching a dolphin to splash. According to her rule, that's all she should focus on—not how big a splash, not the direction of the splash, only splashing. If she trains more than that, she'll confuse the animal. The dolphin will think, Do you want splashing, or splashing to my left? Only once the dolphin's fin hits the water reliably on cue would Pryor then move on to direction.

Take Rosie's learning to ride the skateboard. The student taught her to sit on a board. Only once the baboon did that consistently did he move to the next step, pushing the board with his hand while Rosie sat on it. If he had tried to teach her both at once, she, too, like the dolphin, would likely get confused. "Do you want me to sit on the skateboard, or only sit on it when you move it?"

With humans I thought less about the successive aspect and more about Pryor's directive to focus on one aspect of a behavior at a time. That meant that I be absolutely clear what I wanted to reinforce—what the criterion was, as trainers say. If I wanted my mother to sign up for a hearing test, that is all I would address. I stopped mentioning hearing aids at all. If I wanted Scott to be dressed and ready on time for a dinner party at our house, I didn't also expect him to have drinks poured. If he did have the bar open, swell, but that was icing on the cake. If anyone sent me a gift, I said thanks because I want presents, lots of them, so I'll reinforce that even if it's a dud, such as the gumball machine my dad gave me when I finished graduate school. He still gave me a gift, which is the criterion. However, I may make sure the giver gets my Christmas list the next holiday season. One year I talked my father into giving me his credit card so I could buy my own present to me from him, a gun-metal trench coat. It was too big an approximation for him. I never got my hands on his card again.

I also stopped raising the bar midbehavior, meaning no more "thanks-butting," as I call it, as in "Thanks for getting the groceries, but you didn't get the exact kind of milk I wanted," or "Thanks for visiting, but I wish you could stay longer." Raising the bar midbehavior may not only confuse people as to what exactly you want from them but can also punish them. In that case, you may lose the behavior (grocery shopping, visiting) you thought you were reinforcing, and the grocery shopping may fall to you and the houseguests may never return.

NEW TANK SYNDROME

Even the most thoroughly trained animal can be distracted by a new setting. When dolphins are moved to different tanks, the marine mammals often come down with a case of temporary amnesia as they absorb their new digs. This is what trainers call new tank syndrome, rule number four.

I saw new tank syndrome the day I sat in on a rehearsal of the annual show they put on at the training school. The macaw wouldn't fly. The beaver missed his cue. The serval froze, twitched one of its big ears, and stared. Though all the animals had been taught their parts, on stage for the very first time, with all the other animals and people, they basically forgot their lines. Though the students might have been frustrated, they were not surprised. The animals were essentially learning a new behavior: how to do the old one in a new setting and under new circumstances. A few more rehearsals later, the animals were flying, sitting, and walking on cue again. Time and familiarity had done the trick.

In the face of new tank syndrome, trainers typically lower the bar, or, as they say, relax the criteria briefly. They expect less until an animal has absorbed the new stimuli. If time doesn't do the trick, they may have to go back to kindergarten. The point, as usual, is to have a reasonable expectation. That way neither the animal nor the trainer becomes unduly discouraged.

You see new tank syndrome with humans all the time. This is why sports teams have a home field advantage, if only slight. The other team is in a new tank, with the unfamiliar crowd, lights, the way the air smells, and may consequently get off to a rough start. When you drive in a new city or country, you may find yourself braking too often or going too fast or forgetting to use your turn signal as you absorb terra incognita. I get new tank syndrome every time I have to use a different computer. My writing suffers until I'm accustomed to the keyboard.

Like a trainer, I learned not to fuss over new tank syndrome, to let time do the job. If I beat myself up, trash my own mood, my writing might be shot for the day rather than just an hour. Likewise, if the coach screams at his players from the sidelines, he might make his team so anxious that the whole game will be lost, rather than just the first few plays. And, from my experience, best to cut any driver in a new place some slack, especially if you are the driver's spouse in the passenger seat and it's the first hours of your vacation. Otherwise, the whole trip might be ruined rather than just the drive to the beach.

TRY SOMETHING DIFFERENT

If one training method is not working, try another. This is rule number five on my list, and it's the one that has meant the most to me in a big-picture kind of way. "There are as many ways to get behaviors as there are trainers to think them up," Pryor writes. It's so simple, so obvious, but so contrary to the way we humans go about so much of life, especially our relationships. We are so inclined to dig in, to deepen the rut. I often have. And when I finally am neck-deep in the rut I throw my hands up in surrender. If a technique isn't working, a trainer would not persist like that, at least a good one wouldn't. And they wouldn't take it as a personal failure or blame it on the animal. They would just think up a plan B.

Now when my training attempts fail with humans, I try making the approximation smaller. I dissect my own behavior, consider how my actions might inadvertently fuel someone else's. I ask, What exactly am I reinforcing? If it's with a friend, family, or student, I might consult a fellow "trainer," my husband or Hannah or Elise, for advice. I consider my timing. I coolly analyze the behavioral headache like a mathematical equation, looking for another way to solve it. Then, like a trainer, I try that other way.

There are times when a different approach eludes me. Then, thinking like a dolphin trainer, I figuratively grab a bucket of fish and pull up a seat to the side of the pool. There I wait and watch for the human animals to do something, anything, I can encourage.

LURING

In the morning, the giraffes won't leave their night quarters. In the evening, the gorillas won't go inside. At any time of day, a rhino hovers endlessly in the doorway between his two enclosures. Moving animals from their night quarters to their daytime exhibits, what is called shifting, is often a huge headache at zoos. Many animals balk, for hours in some cases, frozen by a bad case of "should I stay or should I go." To get the animals to decide, some keepers proffer bananas and carrots—positive reinforcement technically, but in this use called luring or baiting.

Luring is putting the carrot before the horse. It is essentially showing your hand positive-reinforcementwise, saying "You will get this if you do that." It's also promising a reward rather than reinforcing a behavior. Hairsplitting, I know, but there is a difference.

There are several ways to shape behavior, luring being one of the oldest and most widely used. Trainers have used it for centuries. A common way to teach a dog to sit is to hold a tasty morsel right over his head, which prompts him to put his bottom on the floor.

Gunther Gebel-Williams of Ringling Bros. fame reportedly stuck a piece of meat on the end of a stick to get his big cats' attention.

Some trainers aren't keen on luring, for a number of reasons. First, it gives the animal a chance to decide in advance whether the treat is big enough or not. The animal may hold out for a bigger banana, maybe a whole bunch, maybe several bunches. And, as with any mistimed reward—with luring, it's too early—the trainer may teach an unintended behavior. What with all the fruits and vegetables and the keepers making a fuss, a dawdling camel may think dawdling is the ticket. In his ungulate mind, stalling in a cage door equals dinner theater. And so the camel stalls day after day.

I admit to luring humans. I've put dinner on the table to get chatty, balky guests to come in from the living room. I talked Scott into going to an IKEA during the store's opening week by promising him Swedish meatballs and apple cake in the store's restaurant (in the end, neither dish could make up for the mosh pit of shoppers we encountered by the throw pillows). I've lured visitors to my house with promises of beach trips, strawberry picking, and ferry rides. If Maine's weather gods agree, I'm good for it. If they don't, as those cranky bastards so often don't, my guests may leave early and never return.

Luring presents some of the same problems with humans as it does with animals. You can hold out for a bigger or better reinforcement or decide the prize just isn't worth the bother. This goes on with kids all the time. A parent promises a bike for straight A's. Midway through the term, the kid realizes getting all A's is far more work than he expected, that a bike's not worth it, and throws in the towel. My mom posted a chart on the fridge of housework with what each chore would pay. My sister and I went to work whenever we wanted new Barbie doll clothes or a bag of bubble gum. The rates my mom paid (one dollar for washing the car, fifty cents for cleaning the bathroom), however, never inspired my brother. For him, the promise of pocket change could not compete with digging holes and building forts in the woods behind our house. Not much is as reinforcing to a boy as jabbing a spade into the soft earth or the dreams such shoveling inspires, of carving lake-sized swimming pools out of suburban yards, tunneling to faraway lands, or unearthing tarnished treasures buried by like-minded boys centuries ago. If my mom wanted my brother to do chores, she needed to pay him in shovels.

EIGHT

THE LEAST REINFORCING SCENARIO

On yet another glorious day in Southern California, I followed the training students on a field trip to SeaWorld San Diego, where I met a four-hundred-pound baby walrus, watched trainers romp with killer whales, and reached elbow-deep into a shallow pool to stroke the smooth, elastic backs of California bat rays. The day ended at Dolphin Stadium with a rousing, splashy show by the park's assorted marine mammals, including Bubbles, one of the oldest pilot whales in captivity. At the show's peak, the half-dozen animals swam to the pool's edge, turned their backs to the audience, and, in unison, beat their gray flukes on the blue water. This tail-powered churning sent wave after wave of cold seawater over the audience. Water landed in great wallops, dousing people who had remained in their seats though it was obvious what the dolphins were up to. A squad of kids, arms raised, squealing, stormed the pool, charging into the surge. Adults, wet polo shirts plastered to their chests, khaki shorts dripping, clambered up the concrete bleachers as fast as their middle-aged legs could go. I watched the pandemonium from high in the stadium, where the trainer next to me had wisely suggested we sit. I asked her if the animals understood what they were doing. "Oh, yes," she said.

After the show, once the soggy crowd had wandered off, a towheaded dolphin trainer, her wet suit still dripping, joined the students. For the aspiring dolphin trainers in the group, she described the park's arduous swim test, offered tips on how to get a coveted job working with the cetaceans, and explained how to use a Least Reinforcing Scenario (LRS), which is what caught my attention. As she put it, when a dolphin does something wrong, say, squirts water when asked to wave a pectoral fin, the trainer doesn't bat an eye. She stands still for a few beats, remains expressionless, nary a frown nor a sigh, and then resumes training. The idea is that any response, positive or negative, might fuel a behavior. If a behavior provokes no reaction, it typically dies away or, as trainers say, goes extinct. In the margin of my notes I wrote, *Try on Scott!*

Back home, it was only a matter of time before Scott was again tearing around the house in search of misplaced keys or wallet, at which point I gave him an LRS. I didn't respond one iota as I heard him storming from room to room. It took a lot of discipline not to, especially when Dixie crept into the kitchen and stood between my legs. However, the results were immediate and stunning. Scott's temper fell far shy of its usual pitch and then waned like a fast-moving storm.

THE UPSIDE OF DOING NOTHING

The LRS was developed at SeaWorld in the 1980s. The four parks have used this method on hundreds of species and on all ages of animals since, from creaky polar bears to baby killer whales. An LRS is, at its core, ignoring behavior you don't want, but in a very specific way. It is used during training sessions and for a brief amount of time. SeaWorld trainers came up with the technique to communicate to an animal when its behavior was incorrect without accidentally encouraging the wrong response. If a trainer asks for a flip but gets a jump instead, he will respond with an LRS, which goes one step beyond not reinforcing the behavior. It says as coolly as possible, "Wrong."

That tells an animal not only when it has made a mistake but also that nothing bad will happen if it errs. In fact, nothing at all happens. In that way, the LRS demonstrates that an animal has nothing to lose by trying. The technique is also used to teach an animal how to behave after a flub. If, rather than swimming off in a huff, a dolphin remains calm during an LRS, a trainer can reinforce that with a slippery squid.

An LRS sounds like a mini time-out, but it isn't exactly. An LRS does not stop a training session. A time-out does, which can be a problem. A clever animal could provoke one to get out of school. Offer a wrong behavior, and bingo, the trainer leaves and the animal can goof off. Generations of high school students have done likewise to get out of history, chemistry, English, whatever class—act up and off you go to detention. Time-outs are also seductive for the human and thus can be overused. They offer an easy solution to a misbehaving animal—just walk away, a tempting escape hatch for a frustrated trainer. Time-outs can also be misused as punishment. Time-outs are ultimately much more of a response than an LRS, and as such, time-outs may encourage a behavior. An LRS, the Switzerland of reactions, is closer to the ideal.

However, an LRS is not perfect. Using the technique does not get immediate results the way correcting or punishing an animal can. An LRS does not in itself teach the animal the correct response. The trainer still has to analyze why an animal is not jumping, flipping, or shaking hands. Because of its very neutrality, the LRS does not work in some cases, notably with self-reinforcing behavior. If a dolphin is having a grand time pushing a ball around the pool or splashing the trainer, an LRS won't have much effect. And standing still for a few seconds is no help when a lion swipes at you. At that point, it's best to move, as in backing toward the cage door.

DO TRY THIS AT HOME

I began using the LRS with all kinds of people. When a veteran clerk at my local post office, a pale woman with round, thick glasses that made her eyes look big and rheumy as a grouper's, snapped at me for not labeling a package correctly, I didn't apologize nor snap back nor try to improve her mood with a smile. I just blankly, quietly fixed the label. This seemed to throw the clerk, who took my money, and then, as she handed back the change, said, albeit without looking at me, "Have a nice day." Only then did I look her in the eye and smile. "You too," I said.

I have some friends who are forever suggesting I get

acupuncture. Should I mention the smallest ailment, from back pimples to asthma, they describe at length how becoming a human pincushion will cure my problem. Thing is, I remind them, I have a terrible needle phobia. I don't get any shot I don't absolutely have to. I grow woozy from a blood draw. The few times I've had surgery I've fretted far more over the coming IV than the surgeon's knife. My protests would only make the pro-acupuncture forces redouble their efforts. Finally, I tried an LRS whenever the subject of long, skinny needles came up. Their lobbying efforts did not cease but lost their typical zeal and breadth. And the pregnant pause of my LRS usually provided a chance to change the subject.

As I used LRSs, I noticed people giving them all over the place, though they did not realize it. Driving through Boston one afternoon just after rush hour, I stopped in yet another endless line of cars merging off a highway. In my rearview mirror, I watched as a blue car with a lone male driver jumped the line and then tried to cut back in. None of the drivers yielded an inch, keeping bumper to bumper so the interloper could not squeeze in. Drivers stared straight ahead, careful not to acknowledge the cad. Everyone was giving him a kind of LRS. When I pulled even with him, he, growing desperate for a response, raged, "Would you *just* let me *in*?" Now he was doing two things I didn't like, trying to cut in and yelling at me. I joined the group training exercise and utterly ignored him, though both my dogs in the backseat turned their heads to see what he was barking about.

The LRS was the most effective technique of any I learned from animal trainers. And I wasn't using it exactly. Most of my LRSs were way too long, and some even defaulted to time-outs. When I couldn't ignore someone or at least remain neutral, I ran for it, as I did with a student. In the middle of a conversation, about the time she burst into tears over an assignment and accused me of ruining her life, I realized that even my reasoned response was too much attention. I announced as matter-of-factly as I could, "It's due next week. Gotta run." I swept up my stack of papers, grabbed my purse, and fled. Maybe not the most graceful or professional LRS or exit, and, to my chagrin, I left behind half a frozen coffee drink in my haste, but by staying I would have, as I had before, reinforced her hysterics.

This method helped me ignore behavior I didn't want, which is no easy task. The human urge to respond runs deep. Not doing so is counterintuitive. We are social animals, after all. Trainers struggle with this human urge when they work with animals. Resisting the knee-jerk reflex to respond to people ratchets up the difficulty a notch. The few beats of an LRS were an approximation of sorts for me. I could almost always manage at least that. And in those few moments I found I could collect myself, think about my next move, or muster the self-control to continue to ignore someone or remain calm. The LRS crystallized for me how any kind of response could fuel a behavior, because you can never predict what will be reinforcing to someone else. It also got me out a behavioral rut I'd been in nearly my whole life.

When I was a kid, my brother Andy, four years my junior, likewise tall and blond, though blue-eyed and with a little pot-belly, had a ferocious temper. I was often the target of his explosions, especially in the heat of neighborhood games of basketball or kick-the-can or when I babysat him. His face would grow red and swollen as a fresh blister. He'd raise his fists, lower his head, and charge me. Any herd animal would have run for its life, but not me. Rather, my instincts were always to hold my ground at the minimum, fight back if need be. I persisted, though he knocked me out with a two-by-four once, hit me on the shoulder blade with a croquet stake so hard it raised a white welt, and threw a handful of silverware at my head (okay, my internal zebra kicked in and I did run that time). The oldest of four, I was the one who got in trouble for our fracases. "I don't care who started it" was my mother's refrain. "You're old enough to know better." Maybe, but was I wise enough? Obviously not.

My brother learned that having a big, fat temper was good fun. You get to clobber your older sister and then watch her get into trouble. I reliably made it even more fun for him by fighting back. I have no idea what was reinforcing about these brawls to me, except that it proved that I was tough, which means a lot to a tomboy, especially one who has undercut her credentials with her Barbie doll obsession. And there was a fair amount of feeling slighted and misunderstood by my mother, which was good material for me to feel sorry for myself. The pity party is so self-reinforcing.

Later it turned out that my brother, who was always given a desk at the back of the classroom because he was so tall, needed glasses badly. Also, his adenoids had swelled so he couldn't hear very well. No wonder the Helen Keller–style fits. Glasses were prescribed. Tonsils and adenoids were snipped.

Still we fought on; our last altercation, the flying-silverware episode, happened when I was a senior in high school and he was in eighth grade. We were probably both what trainers would call "patterned," meaning we'd done something so much it became a reflex.

Now, looking back, I see that all that early fighting patterned me—to stand up to anyone's rage or bullying—for years to come. It served me well at times, notably when a man tried to attack me in a public restroom (I fought back, and escaped), and other times not. When a drunken customer screamed obscenities at the male bartender, it was I, collecting a tray of drinks, who asked him to pipe down. Then he screamed really ugly obscenities at me, the kind men reserve for women, the kind that make people gasp. Not me. "Ha, ha, ha," I laughed just to show how he didn't scare me. He stormed off, found the manager, and complained about me. Then the manager reprimanded me for being disrespectful. As badly as that went, I was undeterred.

A copy editor at a newspaper regularly picked strange, circular arguments with me, often about a point in a story of mine he was editing. He never yelled, but he verbally bobbed and weaved until he had me against the ropes. I knew he loved a good verbal tangle, and yet I couldn't resist throwing a punch. I'd often lose track of what we even disagreed about, yet would hang in there just to make it clear he couldn't verbally or intellectually push me around. I wasted hours arguing with this man.

At home, whenever my husband's temper flared, I jumped, even if his pique had nothing to do with me. It rarely did.

Scott's temper is typically tripped by inanimate objects, such as a panini grill that won't open, an uncooperative computer, and, of course, anything lost. (On the other hand, when I inadvertently threw a frozen water bottle through the rear window of our VW, he was sweet about it.) With Scott, I didn't always puff my chest and dig in for a fight immediately, as I had with so many other people. I might try to appease him or reason with him. Or I'd try to quash his temper by yelling back at him, and a few times we got to shouting so loudly that Dixie would shake. My point is, I always responded somehow, and my responding fueled, or reinforced, his outburst, as it had all the others.

When nothing else had, animal training got me to stop. I finally realized how I contributed to fights with my brother, my coworker, and my husband. I might have gone on with this behavior the rest of my life. My marriage certainly would have suffered. I might finally have stood up to the wrong irate stranger and gotten punched in the nose or worse.

Giving an LRS, ignoring behavior I don't like, demands self-control, buckets of it. I cannot give one in a pique. It is not the silent treatment nor a cold shoulder. Rather, an LRS is a head-to-toe poker face. It demands that at times I not be helpful or sympathetic—maybe, in some people's eyes, not even polite. More than anything it means that I have to hold my tongue, which does not come naturally to me, a fundamentally highly communicative animal. But when I pull off an LRS, tempers wane, complaining fizzles, and pushiness abates. The positive results make having that much self-control utterly self-reinforcing.

LOSING BEHAVIORS

Animals can lose a behavior if they go too long without doing it or being rewarded for it. If a jump or a wave produces nada, the animal thinks, "Why bother?" The behavior gets rusty and then seemingly vanishes. At the training school, Rosie the baboon lost how to do a backflip on a balance beam. Nick the miniature horse lost how to run in a circle on a long lead, what's called "lunging."

Professional trainers call this "extinction" (meaning the behavior, not the animal). That is, perhaps, a bit of a misnomer. Karen Pryor thinks animals never completely forget the things they are taught, but if there's nothing to be gained by it, they shelve the behavior in the far back of their brains next to "Found a tasty seed under a rock by the oak tree five years ago." So once you train a killer whale to breach, a bear to rise up on his back legs, or a parrot to roller-skate on command, you have to practice with rewards occasionally.

This is what is called maintaining behaviors. With animals like Schmoo, the sea lion at the training school who knows some two hundred commands, maintaining behaviors is a full-time job. The student trainers who worked with the dowager spent as much, if not more, time practicing behaviors Schmoo knew than teaching her new ones.

The point is that behaviors—except for self-reinforcing ones—*will* vanish for all intents and purposes if they are not reinforced enough. If you don't want a behavior, that's dandy. If you do, it's not. So whatever you want to keep, be sure to reward it, no matter how big a brain your animal has.

We maintain good behaviors in each other with many of our

customs and manners. Get a compliment, be sure to say thank you. Get a dinner invite, be sure to RSVP promptly and call up the next day to say thanks. Someone makes a friendly gesture, smiles at you, or reaches out to shake your hand, smile or offer your hand back. Miss Manners and Emily Post are, in their own way, animal trainers.

Generally, if you're taking anything for granted, in a friend, employee, or relative, that behavior could go AWOL any day. When someone quits taking out the garbage, setting the table, making the bed like they always have, chances are they were not being reinforced. Chores are rarely rewarded; thus the name—chores—and thus the many fights over why someone isn't doing them.

These days, if a friend or relative stops phoning, I think, Did I reinforce their calls or not? Did I phone back? Was I happy to hear from them, or did I answer distractedly while watching a television show (something my father used to do to me and the reason why I quit calling him)? I, like every other human, miss many chances to maintain behaviors because I'm preoccupied or just plain lazy.

Here are some behaviors I myself have lost over time: reading novels, entering cooking contests, and being in love with a boyfriend. Having gotten nada for those behaviors, I shelved them in the far back of my brain next to "Found a pair of size 9 shoes in a size 8 box at a fabulous sale twenty years ago."

Reading novels was once a self-reinforcing behavior for me. It was my favorite way to kick back, especially in my early days of freelance writing, when I paid the bills by waitressing. After six-hour shifts of nonstop juggling of drinks, orders, and customers, I'd quiet my body and brain by sinking into a chair under the weight of a Dickens or Hardy novel in my lap. But the more I spent my days sitting on my rump agonizing over word choice and sentence structure, the less relaxing I found settling in with a thick book in the evening. Then I began to feel guilty about not reading, which made it even less reinforcing. In the end, I lost novel reading, but, happily, gained another relaxing behavior—dog walking (from my pooches' point of view, patrolling our territory).

Inspired by my first book about competitive cooking, I entered about half a dozen contests and won exactly none of them. Even my brilliant halibut poached in grapefruit juice with couscous was a no-show. As a seasoned cookoff champ told me, you need "a few wins" for the habit to stick. From what I saw on the cookoff circuit, one win could keep someone at it for years. Having gotten nary a pellet of reinforcement, I got mad and hung up my apron. My competitive cooking behavior went extinct. No one noticed.

In my early twenties I fell madly in love with a young man with deep-set eyes and a baritone voice that easily cracked. In short order, he and his thousands of record albums moved in. We planned a long trip to Europe and looked to a future together. And then, about six months into our romance, the love of my life began to ignore me. He quit holding my hand. When he met me at a train station, he didn't stand up to hug me. He would walk ahead of me on the sidewalk.

He was still funny, handsome, and smart, but with so little reinforcement, my feelings faded. He did not maintain the behavior of my adoring him, so I lost it. I set that behavior on that shelf at the far back of my brain, and I broke up with him. To my surprise, he cried and said he had thought we would get married. Regardless of

> what Pryor says, this animal could not recall for her life how to love this man. That behavior had truly become as extinct as the dodo bird.
>
> Lucky for Scott. One person's lost behavior can be another's gain.

Professionals talk of animals that understand training so well they eventually use it back on the trainer. A chimpanzee that had been trained at the National Zoo reportedly handed his keeper a piece of celery after she opened a door to his yard. I heard another story of a dolphin that swam to the far side of the pool when its trainer rewarded it for a behavior with the wrong kind of fish. The dolphin would pause there for a few moments, then return to the trainer. The trainer realized the dolphin was giving *her* an LRS, figured out her mistake, and went and got a bucket of the right kind of fish.

One of my animals did the same to me.

While I worked on my previous book, every time I returned home from the school in California, I chattered on and on about training, eventually talking about how I was using the ideas with humans. I talked to no one about this more than Scott, the subject of so many of my experiments. He wasn't offended, just amused. As I explained the techniques and terminology, he soaked it up. Far more than I realized.

As I was finishing my previous book, I awoke one morning to find that my mouth, like a jammed door, would hardly open. Nothing hurt, but something was clearly wrong. Afraid to eat but starving, I cut a banana in half lengthwise and threaded it through my only slightly parted teeth. That was the only way I could eat a banana for months to come. I gave up sandwiches, including hamburgers, and maki rolls, neither of which would fit in my mouth. Carefree days of eating French bread, caramels, corn on the cob, and smoked almonds came to an end.

A line of muscles from my left shoulder to my cheeks had inexplicably tightened. The cause was hard to pinpoint, but certainly months of hunching over a laptop had not helped. An orthodontist blamed my faulty jaw on my faulty bite. So, firmly in middle age, desperate to yawn with gusto once again, I got braces. They were not only humiliating but excruciatingly painful. For weeks my gums, teeth, jaw, and sinuses throbbed. I complained frequently and loudly. Scott assured me that I would get used to all the metal in my mouth. I did not.

One morning, as I launched into yet another tirade about how uncomfortable I was, Scott just looked at me blankly. He didn't say a word or acknowledge my rant in any way, not even with a nod.

I quickly ran out of steam and started to walk away. Then the lightbulb went on, and I turned back to him. "Are you giving me an LRS?"

Silence.

"You are, aren't you?"

He finally smiled, but his LRS had already done the trick. The animal had begun to train the trainer.

NINE

THE JOY OF INCOMPATIBLE BEHAVIORS

The African crowned cranes at SeaWorld San Diego developed an annoying habit—landing on the trainers. As the trainers strolled through the animal park, the cranes, their great white and slate-gray wings arced against the wind, would fly overhead. The cranes came when called but would alight, wings flapping, dinosaurlike feet extended, on the trainers' shoulders and heads. It was like having a golf umbrella tumble down on you from the heavens. Though the cranes weigh only six to eight pounds, the birds can be a leggy four feet tall. They have a long hind toe, which allows them to roost in trees but also to pinch shoulders and grab hair. You really don't want a crowned crane to roost on you.

Rather than try to stop the cranes from landing on them, the trainers taught the birds a behavior that would make the undesirable one, balancing on their heads and shoulders, impossible. The trainers taught the birds to touch down on mats they tossed on the ground in front of them. The birds couldn't alight on the mats and their heads simultaneously.

During my visit to the park, a male trainer, tan, fit, and bleached blond as a lifeguard, turned and strode across a plush lawn to demonstrate. A crane in the air, its signature spray of gold atop its head glimmering in the noontime sun, took note, unfolded its great wings, and, with a flap or two, pulled skyward like a kite, long legs trailing behind it. The trainer turned, called the bird, and pitched the mat in front of him. The crane tucked his wings, stretched his legs out, spread his toes, and touched down neatly on the mat.

This is an example of what trainers call an "incompatible behavior," a simple but brilliant concept. The essential premise of the incompatible behavior is to teach an animal to do something else rather than stop something. The twist is that trainers pick another action that makes the first impossible to do. If Shamu pesters his poolmates, have him present his fluke or pectoral fin poolside. He can't do that and chase the other killer whales around.

Trainers use incompatible behaviors all the time. At the training school, whenever Rosie the olive baboon blinks rapid-fire at someone, which is baboon for "I want a piece of you," the student trainers run her through what they call her "control behaviors." The olive baboon mimics the trainer as he covers his eyes, his ears, and his mouth, the old see-no-evil drill, which is very appropriate given the context. However, as cute as the pantomime looks, the point is to make Rosie concentrate on her trainer. She can't do that and shoot dirty looks at someone else at the same time.

Holed up in a windowless conference room in Baltimore, I and a roomful of zookeepers listened to how two aquarists had solved the conundrum of Lance and Dottie with an incompatible behavior. The two spotted eagle rays at the Living Seas, Disney's 5.7-million-gallon aquarium in Florida, had become a bother. The fish have a long, whiplike tail and can grow to have a wingspan of ten feet, though Lance and Dottie were, luckily, not that big. Whenever divers dipped into the tank to feed the fish, the two rays nipped at them, tugged at their masks, even yanked off their regulators. Rays don't have teeth, but they do have a bony palate they use to crush shellfish. That shell-crushing palate can really pinch. Swimming in the tank became less and less reinforcing, as they say, for the divers. The divers didn't even have to be in the water to get pinched. In a video the trainers showed at the conference, a female diver sits on the tank's edge, her behind slightly overhanging the water. The water riffles near the diver's bottom. She jumps, cries "Ouch!" One of the rays had just goosed her.

The pushy behavior likely resulted from the divers' carrying the rays' dinner on their person and feeding the rays by hand. Consequently, the rays got to thinking of the divers as human food dispensers. A training solution was in order, but as far as anyone knew, no one had trained spotted eagle rays before. The bruised aquarists had nothing to lose by trying.

They decided to teach Dottie and Lance an incompatible behavior. They put the rays' dinner in a little PVC elbow, which they fixed with a lid that the rays could open. Then the divers held the improvised feeder in their hands, an arm's length away from their bodies. The rays couldn't harass the divers and feed at the same time.

It took some time, months in fact, but the rays caught on. As they nudged the feeder with their wide, flat snouts, they left the divers to paddle in peace. The video ended as one of the rays glided through the tank's gray-green light toward an empty-handed diver. That diver pointed to another, the one holding the feeder. The ray U-turned, fluttered its wide dark wings, and sailed off toward the diver with his lunch.

———

The beauty of incompatible behaviors, it struck me, is that they tap a natural phenomenon: that it takes more energy to stop a moving object than to change its direction, meaning it may be easier to get someone to do something else than to stop them from doing something. The technique requires some creativity, even a sense of fun. It also pays to know your species, because you can easily shape instincts, or things your animals naturally like to do, into incompatible behaviors.

It first occurred to me to try one with Dixie. Whenever I loaded the two pups into the backseat of my car, Dixie, the alpha, felt compelled to remind Penny Jane who was boss. She'd block Penny Jane's way, or hump her. I tried to stop Dixie's pushiness with little success. Getting into the car was not much fun for Penny Jane.

So I began to put Penny Jane in the car first, and then Dixie. The moment Dixie's paws hit the backseat, I'd ask for a kiss. That dog loves kissing, so it's no problem to get her to pucker up, and she couldn't lay one of her signature humongous licks on my cheek—she often gets her tongue in my ear somehow—and push Penny Jane around at the same time. Kissing me also seemed to take Dixie's mind off strutting her alpha stuff. Then she would just take a seat, and off we'd go.

That problem solved, I turned to another member of my pack, the one known as Big Dog. Scott had a habit of hemming me in at the stove when I cooked. This annoyed me. I'd ask him to give me some space, and he would, but before long he would drift back just as I was folding an omelet or pouring wine into a hot pan. I'd bark at him, and he'd seem surprised to find himself by the burners again. Flames, food, his mate—I think he couldn't help himself. He instinctively drew close.

To lure him away from me and the stove, I piled up parsley for him to chop or cheese for him to grate at the other end of the kitchen island. Or I'd set out a bowl of chips, salsa, and a beer across the room. Soon I'd done it: My husband could not be in two places at once. An incompatible behavior tailored to Scott's natural inclinations—he is a food-driven, actually chip-driven, animal, after all—had done what years of barking had not.

Inspired by my success, I tried one on my mother. When I visit her, I like to buy her a thing or two for her house. The idea is that this would be fun, that we'd shop together and she'd pick something out herself. But most times, once in the store, she would launch into a laundry list of reasons why she didn't need anything, a litany that would often follow me to the cashier. It was as much fun as having a flapping crane come in for a landing on my head. My next visit home, I scanned her small ranch house, decided on what I thought she could use (nearly always lamps), asked to borrow her car, made up a reason why, and shopped by myself. My mother's being at home was incompatible with her obstructing me from buying her a gift. I arrived home with a standing lamp in one hand, a table lamp in the other. "Oh," she said. "Well, you shouldn't have." Then we had some fun figuring out where they should go.

I thought of an incompatible behavior, actually more of an incompatible situation, for a neighbor who'd gotten to thinking of our condo building's scrubby backyard as his own. He stored bags of mulch, old skateboards, plastic bins, and an unwanted teak table on our side of the fence. One fellow condo owner wanted to read him the riot act. I doubted ultimatums would go over well, and seeing he was a neighbor, it was important to keep good relations with him. Plus, though it was our backyard, he'd been stashing stuff in it for years without anyone's saying boo. As a trainer would say, he was patterned.

I decided to landscape the yard. As I added rhododendrons and hydrangeas, there would be less and less room for his stuff. He'd have to move it. The backyard needed some TLC to begin with, and I figured this neighbor, whose house overlooked our dismal plot, would appreciate any improvement. It would take time, but I thought we had a better shot at results this way than by confronting him.

I thought of one for my student prone to hysterics. As I mentioned earlier, I gave her an impromptu time-out once, but I needed a more thoughtful solution. I realized she saved her emotional fireworks for when we were alone together. Given that, the incompatible behavior was obvious: Talk to her only with other people around. I guessed, correctly, that this would keep her rational. This solution was like asking a killer whale to present its pectoral fin poolside. The behavior keeps Shamu from bothering his poolmates, but the pectoral presentation in itself, a quiet kind of action, promotes calm.

It was tricky with the student, because she would linger after class, stand back while other students peppered me with questions, and approach only once everyone had left. So after class, as students clustered around me, if I saw my high-strung one dawdling, I'd turn to her. "Do you have a question?" I'd ask, making the circle of students wait and thus keeping them in the room. While she and I talked, if I noticed the rest of the class drifting toward the door, I followed along. If she wanted to keep talking to me, she had to follow me out on the street into a great vortex of human animals, where we were sure not to be alone.

I tried a few incompatible behaviors on myself too. I hate waiting, and as the aimless seconds pile up, I grow crankier and crankier. Given that I run at higher rpm than my husband, I would typically find myself killing time while he showered, shaved, dressed, and did God knows what else for a night out. I needed a behavior incompatible with waiting. Now, whenever I'm ready ahead of him, I weed my garden. I like weeding. It combines exercise, fresh air, tidiness, nurturing, and dominance (I decide what will live or die) all in one simple act. Weeding not only keeps me busy, and thus technically is an incompatible behavior to waiting, but the physical act of it, yanking the life out of unsuspecting plants, keeps frustration at bay. This has two drawbacks. Weeding is dirty work. I have opened a menu in a swank restaurant and gasped, not because of the entrée prices but from suddenly noticing the black loam jammed under my nails. I've sat down in the dark of a movie theater only to notice a strange, nearly phosphorescent orange smear on my pants, which in the light of the bathroom I identify as daylily pollen. Weeding is also a seasonal incompatible behavior. During New England's many barren months, I might straighten up the kitchen instead, but that's not near the fun weeding is. In fact, it's really not fun at all for me. So though I'm busy and productive while cleaning, a counterful of smudged wineglasses can spike my frustration in a way mere waiting wouldn't. Not all incompatible behaviors are created equal. I need a winter plan B.

———

Incompatible behaviors have been around at least as long as there have been battle plans. If you draw your enemy to a battle line, that is an incompatible behavior to having their forces attack you from the rear. Lots of people—teachers, coaches, bosses, parents, spouses—use this technique. When I went to city hall recently to renew my dogs' licenses, I found that the pen on the counter had sprouted a tail. A long plastic spoon was taped to the end. The clerks had done so to keep distracted people such as myself from walking off with it. Making the pen essentially twice as long was an incompatible behavior. People can't absentmindedly pocket what doesn't fit in their pocket. They could have put up a sign that read "Don't Take the Pen or Else!" but the incompatible behavior was a surer, not to mention nicer, way to keep the writing instrument in its place.

Sending someone on an errand to get them out of the way is an incompatible behavior. Each holiday season, I notice many more men in the grocery store. They wander aimlessly with crumpled lists clutched in their big hands, looking totally at sea amid the bright lights and orderly shelves. It's like some strange annual migration. Now I realize why they are there, looking for mayonnaise among jars of jelly, contemplating gleaming shelves of olive oil, kneeling to scan plump cans of tomatoes (crushed, whole, peeled?), growing woozy from so many choices. The cook at home may need a few items, but I suspect that is rarely simply the case. Their wives have given them an incompatible behavior. To get them out of the house while they prep the holiday feast, the wives send their husbands to fetch sugar, butter, cranberries, or, if they want to be sure they'll be gone for a while, hard-to-find items like cheesecloth, capers, and a packet of yeast.

Parents especially make use of this approach, not realizing it has a name and a parallel in the animal training world. My mother rarely quieted us down. Rather, she'd send us outdoors to play, where we would not raise the decibels in the house. "It's not raining that hard," she'd say as she helped us find our slickers. A lot of mothers in our neighborhood had the same idea, so we had lots of company when playing in the rain, snow, or whatever. The bonus was we grew up to be a very hardy lot.

I had used a few incompatible behaviors in my day, long before I knew that's what they were. Anyone who's ever babysat knows how hard it can be to get your young charges into bed. Most kids love having babysitters, and the break in routine whips them into a frenzy. In my early days of babysitting, as the car lights flashed across the house announcing that the parents would come through the front door any second, I got on my knees and begged three grade-schoolers to get in their beds. Finally, I got smart and began wearing the kids out to the point that they'd willingly go to bed. Utter exhaustion is incompatible with refusing to hit the hay. I came up with wild, physical games. One of my most successful was the Loud Restaurant. I, the customer, and two little boys, the waiter and the cook, yelled back and forth at each other. "I'd like a hamburger!" I'd bellow. "How would you like it cooked?!" my mini-waiter hollered back. The mini-cook would bang pots together and scream "Ready!" After an evening at the Loud Restaurant, I had to carry those kids to bed.

Incompatible behaviors are just such a nice way of doing business, so nice they can be very reinforcing to the giver as well. It was much more fun to set out chips and salsa for my husband than to ask him over and over to back off. I enjoyed the simple pleasure of buying my mother a surprise. I liked gardening our back plot far more than I would have confronting my neighbor. The Loud Restaurant was fun for me too. And getting a kiss from Dixie dog always makes my day.

Now, as each figurative crane, wings flapping, feet out, tries to land on my head, I reach for a mat.

TEN

WORKING WITH BIG CATS

Whenever Mara Rodriguez walks one of the two cougars around the training school on a chain or steps into their shared cage, she watches for telltale bits of body language that broadcast that the cats are beginning to feel like what they are—serious predators. She is on the lookout for dilated pupils, a head hung low, a slight crouch, or a stare that can't be broken, a kind of predatory trance. She doesn't want a cat locking its eyes on her specifically, but its staring at anything could spell trouble. A cougar peering at a tree may have spotted a rabbit or other snack hiding behind it. If the cat gives chase, the trainer on the other end of the chain will go along for the ride.

Once as Rodriguez walked along with a student leading a cougar on a chain, she noticed the slightest change in the cat's pace. She took the chain. About ten steps later the big cat looked behind him, and, for no obvious reason, bolted. Rodriguez clung to the chain, dug her knees in, and then went flat on her belly to make herself as hard to drag as possible. He towed her about a dozen feet down a gravel road before he stopped, tearing Rodriguez's shirt and pants and raking the skin off her forearms.

Some bad behavior cannot be ignored: a lion lunging, a parrot biting, even a deer rising up on its hind legs (watch out for those front hooves). When a set of fangs is aimed at your jugular, you cannot feign indifference so as not to reinforce the desperately unwanted behavior. You are well beyond the reach of treats. An LRS will get you killed. Better to have a crop, a cane, or a bop stick in your hand, though a trainer recently got away from an irked Shamu, who had grabbed him by the foot, with nothing but his training chops. He remained calm and did not struggle until Shamu let go.

Ideally, I learned from trainers like Rodriguez, you never let it come to that dreadful point where tooth, claw, or hoof meets flesh. Rather, you watch for the subtlest, earliest signals that an animal is heading in a direction you'd rather it not. Then you cut the behavior off at the pass. From some reports, that may have been what Roy Horn was doing the night his tiger attacked him onstage in Las Vegas: The cat had fixed its eyes on someone in the audience. Horn thunked the cat on the nose to distract him and break the stare. Then, unfortunately, the cat grabbed his arm, and in the following tussle Horn fell to the stage, where the cat bit him on the neck.

If Rodriguez sees even the smallest sign, what trainers call an antecedent, that a cougar might get aggressive, she redirects it. That is, she distracts the cat, with a command, a change of direction if she's walking it, or even a pop on the nose, as if to say, "Snap out of it, pal!" Or she leaves the cage. She does something immediately to stop the behavior before it gains any traction. You may be able to distract a big cat in those first few seconds after it's just begun *thinking* about attacking, but not once its feline mind is made up.

Most species broadcast their intentions and moods. Doing so may persuade some pesky juvenile to cool it, send an unwanted Romeo packing, or settle a territorial dispute, all without a scrap. Peevish dolphins and killer whales slap the water with their pectoral fins, squirt water, blow bubbles, or pop their jaws. An agitated sea lion may swing his head or abruptly change his expression. Parrots raise their head feathers, thrust their beaks, or growl. Baboons bark or blink exaggeratedly, what is called eye flashing.

Any abnormal behavior can be a sign as well. Trainers take note of a nocturnal animal wide awake at dawn, a usually restless animal lollygagging, a gregarious one suddenly overcome with shyness. Such out-of-the-ordinary conduct may communicate a mood shift for the worse, not to mention a serious medical problem. In general, an unbroken, big-pupiled stare is rarely a good sign with most species, especially when directed at you.

Individual animals may have their own unique signs—yet another reason trainers must know their charges. As I mentioned earlier, the dromedary Kaleb was prone to hissy fits, big ones, during strolls around the school. Students working with the animal had to learn to read his cues. If Kaleb's state of mind slid south, the thousand-pound-plus ungulate would raise his shaggy tail, tense his large body, push his rounded ears back, and retract his lips, maybe even foam a little from his big mouth. If this body language seems awfully obvious, consider that Kaleb didn't like the person holding his lead to look at him. Doing so could set the camel off. One crack student learned to size up Kaleb's temper by looking at his shadow. Eventually, she knew his body language so well she didn't need to see it at all. She could feel the storm gathering in Kaleb through the lead. When she did, she'd pull down hard on the rope to lower his big head. If that move didn't redirect his mood, she at least would have better control of the soon-to-be-seething camel.

READING THE CUES

Body language is nothing new to me. Like most humans, I use and read it, admittedly poorly at times. Whenever my father pointed his middle finger at me, one of his bad jokes was imminent. (*Did you hear about the dog who drank gasoline? He ran around in circles and then suddenly collapsed. He ran out of gas!*) When the alcoholic–compulsive gambler manager of a restaurant I worked in emerged from his lair of an office in the basement, yelling was in the offing. The argumentative copy editor at the newspaper would stare at me before he picked a fight. A poker club member always asked to look at a cheat sheet whenever she had a good hand. For someone who could never remember if a full house beats two pairs, she won a lot. Luckily, she had that big, fat tell. As soon as she peeked at the cheat sheet I would fold, and thus redirect her behavior ever so slightly. She'd still win, but not as much from me.

I once worked for a newspaper editor in Vermont who, every time he got a story idea, would stand up, stretch his arms showily, let out an exaggerated sigh, rub his belly, and feign pitching a baseball. Then he would scan the newsroom for an idle reporter. You didn't have to work at this paper long to know that the minute the editor stood up and stretched, and definitely by the time he patted his stomach, you wanted to look very busy. We seasoned reporters would pick up our phones and pretend we were intently interviewing someone, pounding away at our keyboards as if we were taking notes.

In trainer talk, I was reading this editor's cues that preceded a behavior I didn't like: the delegation of another assignment. I had enough work already, so I redirected his behavior away from me and onto some poor schmuck reading *The New York Times* at his desk. This, however, was an exception. More often than not, I didn't respond to cues, even big ones like the editor's, until it was too late, if then.

I, like a lot of people, tended to ignore all kinds of signs, subtle and not, sometimes out of politeness, sometimes out of denial, until they escalated to a point that I'd have to do something. I would hope the editor would turn his gaze elsewhere, that the manager wouldn't scream at me, that my father might finally tell a good joke, not the one about the gas-guzzling dog again. An acquaintance told me he thought people ignore cues because hope springs eternal in our bosoms, that this time will be different, that an overtired child won't erupt, that a friend won't whine, that a boss won't yell at you. Maybe, but there's a fine line between hope and laziness. I know a heart full of hope has often led me right down the path of least resistance, which often leads the wrong way.

When I was an au pair in France one summer, whenever I grew tired of my three charges, I would just turn off the French-speaking switch in my brain. Then they could talk all they wanted at me, and all I heard was a blissful "Blah, blah, blah." During one such episode we were eating dinner. I spread some soft cheese on a slice of baguette, and as I lifted it to my mouth, Alix, the five-year-old, "blah, blah, blahed" at me, more emphatically as the bread got closer to my mouth. "Blah, blah, blah!" She waved her little hands and pointed at my snack. Finally, as I opened my mouth, I saw that the morsel in my hand was moving. The cheese was, you might say, beyond ripe. I snapped my mouth shut and flicked on the French switch. "There are little animals in your cheese!" I heard Alix cry. I pitched the whole wriggling mess into the yard.

DISPLACED AGGRESSION

Angry, frustrated animals, especially primates and big cats, often strike out at whatever is at hand. Once, during a training session at the school, Kiara the lioness accidentally fell off the shelf in her enclosure and splatted on the concrete floor. Turns out lions aren't all that graceful. She let loose with a toothy, full-throated roar, then swatted a log suspended from the ceiling.

"Get it, go get it!" the student trainer egged her on. Eventually Kiara, having shown the log who's boss, got over her pique, sauntered back to the waiting student, and began to train again.

This is what trainers call displaced aggression. You won't find it just at the zoo. I see it every day, in my rearview mirror as an SUV pulls close to my bumper or from the sales clerk who snaps over a twenty-dollar bill. Spouses of either gender regularly feel the claws of displaced aggression. So do kids, parents, siblings, and employees. My mother had one famous episode of displaced aggression on a family vacation trip when the silverware drawer in our pop-up camper wouldn't open. My mother channeled all the frustration of a rainy, dismal week in the Smoky Mountains, of sleeping bags that wouldn't dry out, fires that wouldn't start, the constant mud on the camper floor, onto that drawer. As she yanked and yanked and yanked, we four kids, who were jammed in the camper because it was raining again, backed away, then pressed flat against the canvas walls. We could have dashed out, but it was like watching some mysterious natural phenomenon. We had to see what would happen. My mom jerked hard one last time and suddenly the drawer came free. Knives, spoons, and forks sailed through the camper and showered down noisily around us. We were less scared than awed at her power, not to mention the flying cutlery. Later, we laughed endlessly about our mother, the silverware bomb.

Displaced aggression is not fair, not right, but a fact of life. Given that, I decided to quit taking it so personally, especially from my husband. Don't get me wrong, I don't like it. Which brings me to my second point. That is why I pointedly avoid it now. When my husband is wrestling with his bike, say, the saddle won't go on right or another greasy gizmo won't cooperate, I do not talk to him, not even to pass on the most fundamental bit of info, such as "Your toast is ready" or even a simple "Hello." I would break this rule only, maybe, to yell "Leopard!"

By the reverse, if I feel like taking a swipe at someone, typically my husband, I look for a log to swat instead. This fall, when I couldn't find a parking sticker we needed for our car, I went to pieces. It had been a bad day, preceded by a bad week. The lost sticker was the last straw, the stuck drawer of silverware. Scott offered to help me look for it, which inexplicably made me want to divorce him on the spot. So with what little self-discipline I could muster, I marched myself upstairs and threw myself on the bed to rant and rave about stupid parking stickers, marriage, life. Somehow I managed all this while also paging through a catalog. Eventually the images of carefully folded comforters and brightly colored sheets calmed me. As my fit of temporary insanity lifted, Scott called upstairs, "I found the sticker."

A trainer cannot afford not to pay attention nor to respond, ever. An overlooked cue, even from a small animal, can have big consequences. Trainers can't stand around hoping everything will work out on its own, that the glaring cat won't jump them, that the agitated monkey won't bite. At the teaching zoo, the novice trainers, given the heady combination of too much confidence and too little experience that comes with being a student, didn't always heed signs that a pro would. One of the best students, who had trained a camel, a baboon, and the cougars, entered the kinkajou's enclosure one morning even though something was obviously up. The kinkajou, a nocturnal rain forest mammal that resembles a fairy-tale creature with its wide shiny eyes, perfectly pink snout, and long prehensile tail, was wide awake and on the floor. Typically, he would be fast asleep in his den box. Regardless, the student stepped inside the cage. In short order, the kinkajou locked onto the student's hand with its small but sharp teeth. There it remained for several minutes, puncturing the student's knuckle and ego.

———

A trainer told me how she used reading cues at home with her son. She observed that right before he got emotional, he would lift an eyebrow. Seeing an eyebrow rise, she'd quickly change the subject, maybe tickle him. The trick was not just in seeing the cue but in jumping on it. It occurred to me that I might do the same with my human animals: start paying attention to cues, and, moreover, like a trainer, respond to them. It might save me from some figurative bites, solve a nagging behavioral conundrum or two. One quickly came to mind.

Most nights, Scott beats me to the bathroom. That done, I can wait a good half hour before I can wash my face as he studies entries in the *Encyclopedia of Rock and Roll*. I've rapped on the door and asked how long he'll be. I'll hear a page turn as he calls out "Just a few more minutes." As in fifteen more minutes. Let's put it this way: The man has read so long on the toilet that his legs have fallen asleep. Ignoring that he was in the bathroom didn't work; neither did getting huffy. I needed to stop the behavior before it started, meaning I had to get to the bathroom first, scrub, slough, floss, and brush. Then he could read from Aerosmith to Frank Zappa without bothering me.

Typically, however, I would be sprawled across the couch with our pups watching *Animal Planet* and considering calling it a night at the next commercial break, when I'd hear the thump of the bathroom door upstairs and realize I'd missed my chance again.

What would an animal trainer do? Respond to the early cues that my husband is headed for the bathroom to read. Those would be a magazine in his hand as he wanders toward the staircase or an end-of-evening request such as "Have you seen my bike catalog?" Seeing or hearing such a cue, I now make a break for it, calling, "All I need is a minute!" Thus redirected from going into the bathroom, Scott reads in the comfort of an upholstered chair, where his legs do not fall asleep.

The biggest breakthrough for me was learning to read and react to my own body's cues. I have mild asthma. I've never rushed to the emergency room; still, even mild asthma is worrisome, because mild can always morph into major asthma, as a number of allergists have lectured me over the years. "Use your inhaler, use your inhaler," they all chanted. I'd nod but go on disregarding all the signs, hoping my lungs would just cool it. Huffing and puffing to keep up with my dogs, I'd finally track down my inhaler. Usually by then my asthma would be so entrenched that I'd have to reach for the big gun, my steroid inhaler, and it would take days, even weeks, to tame my lungs. If my asthma were a lion, I'd be dead.

SCENARIOS

Say my asthma *was* a lion: What was the earliest sign my lungs might aggress? That was when I should redirect my asthma with a couple of puffs on my inhaler. I had no idea. I had to pay attention. The next few times my lungs cramped, I took note. It was the slightest sense of something pushing on my chest.

That is what animal trainers would do. Sometimes, to learn what leads to or provokes a behavior, they have to let the behavior happen. Then they watch for the early signals an animal gives off, the antecedents, so they know what those are next time. Trainers also note the events or circumstances that prompt the behavior, which they call precursors. A common precursor that sets off many captive animals is gating, moving from one quarters or tank to another. Another is being examined by a vet. At the training school, a precursor for Kaleb's fits was the bark mulch trail near the Galápagos tortoise's enclosure. Walking over the loose chips nearly always got the camel kicking and rearing, and in fact that's where student trainers took Kaleb when they wanted to practice controlling his temper tantrums. They'd pull down on his head before they even hit the trail. If a trainer knows what precursors trip an animal's unwanted behavior, he can avoid or change them, or at least be prepared. The trick is to see more than just the behavior, to see what comes before it, what surrounds it, the scenario.

The scenario is everything that leads up to a behavior—all the precursors and antecedents, the whole ball of wax. Scenarios for bad human behavior, from small to epic, abound. Every morning in houses across America, the get-the-kid-to-school-on-time scenario unfolds, starting with innocent questions such as "Do you have your homework?" and peaking with much figurative (or sometimes literal) hissing and scratching. Spouses leaving—to go grocery shopping, out on the town, on vacation—are also scenarios for much aggressive snarling. If I were a divorce lawyer, I think I'd put my office in an airport and take drop-ins.

This is the scenario for bad behavior between me and my

mother: We're eating out in a restaurant, anything from a chili parlor to a five-star French restaurant. We've finished eating. The waiter brings the check. We both reach for the slip of paper, snatch it, pull it back and forth yelling "My treat," "No, my treat." This table rugby has a long tradition in my family. Some of my earliest memories are of my mother, aunt, and grandmother diving for the check, ripping it out of each other's hands. My mom and aunt would whine "Mo-om," even though they were grown women. My grandmother, who typically won these grab fights, would hold the check high and smile deviously. She liked winning anything. This rehearsed aggression, as a trainer might call it, was just part of a family dinner out for me. Not so for Scott, who was mortified by this family habit. He wanted to treat his beloved mother-in-law to dinner, but he didn't want to have to arm-wrestle her for the check.

So Scott and I analyzed the scenario: The check hits a restaurant table and at least two members of my family are seated at it. Who paid for dinner, we realized, had to be decided long before the meal ended. Since I was part of the check-wrestling scenario, I had to withdraw figuratively. Next time we ate out with my mother, as soon as we sat down, my husband told her he'd like to buy her dinner. "Okay," she said. When the waiter brought the bill, I, feeling the urge to grab it, an early sign that I was going to grab it, redirected myself. "I'll get our coats," I said, and rushed off. My mother folded her hands in her lap, an instinctive incompatible behavior to lunging for the check. My husband calmly reached for the bill, and in that moment a behavioral pattern two generations in the making was finally broken.

LIFE AFTER SHAMU

ast fall, Scott and I boarded a ferry to a mitten-shaped island we could see from Portland. Though it was Veterans
Day, and the trees had shed most of their leaves, the sun was
hot and bright like a summer day. It was so hot, we left Dixie
with her heavy coat behind and ventured forth just with
Penny Jane. The island was small enough that we could amble
from one end to the other along narrow, mostly empty roads.
Scott shot hoops with a half-deflated basketball someone had
left at a public court, while I pressed my face to the window of
a vacant summer house. We hunted for sea glass on a gray,
grainy beach while Penny Jane loudly crunched mussels. We
strode along the eroding, vine-covered military fortifications.
We scampered out onto the island's rocky northern tip that
jutted into the bay like a ship's bow.

As the day faded into its final golden hour, we tramped back south to catch the ferry home. At the dock, one of us, I can't remember who, checked the boat schedule that was posted and discovered that the next boat wasn't arriving shortly, as we thought. It wasn't due for two and a half hours, way past dark, way past dinnertime for me, Scott, Penny, or Dixie back at the house. Neither Scott nor I had brought our cell phones, not to mention jackets. That morning, when Scott had checked the schedule on the Internet, he'd forgotten it was a holiday, when fewer boats run than normal. I sighed and kicked at the dock's wooden planks. Scott groaned and stomped.

Normally, the sighing and stomping would have escalated to at least a squabble, and the great day would have ended on an off-key note. But as I felt a recrimination rise in my throat, I thought of the crowned cranes and I reached for a mat. There was a stack of newspapers blessedly on the dock, recent ones even. I grabbed one and sat down in the still-warm sunlight and unfolded it. Perusing it from cover to cover was an incompatible behavior to sniping at my husband (I don't talk when I read) and to waiting idly (I would be busy absorbing the day's news). Scott, whether consciously or unconsciously, gave himself an incompatible behavior too. He stood at the end of the dock and waved his arms wildly every time any kind of boat sped past. I did look up from the paper once to say something grouchy like "No one will stop," and thought to mention how ridiculous he looked flapping his arms so, but then realized Scott had found something to do and left him to it. Penny Jane, her tail curled, her nose down, sniffed a small beach just by the dock.

Twenty minutes later, about the time I was down to the sports section, a woman with frizzy hair and clogs, one of the island's few year-round residents, thunked out onto the dock. Scott lowered his arms. I looked up from the paper. We stared at her in wonderment for a few beats and then asked if there was a boat coming. She said yes, her friend was on it. This boat, she explained, wasn't listed on the schedule. We turned our eyes south to the horizon. Sure enough, in short order we spotted a big white bow plowing our way.

The ferry pushed against the dock. Its few passengers walked off, and then we, rescued from two hungry, cold hours, stepped lightly up the gangplank. Other than the crew, we three had the boat to ourselves. As the engine surged, we hurriedly clanked up metal stairs to the top deck and then stood against the rail as the boat pulled away from the island. In the past, if we had fought on the dock, the trip back would have been a mostly silent, slow rapprochement, with maybe a few generic comments between us about the scenery or Penny Jane. We would have been too sulky to enjoy the boat ride or that everything had worked out. As it was, we huddled to keep warm in the growing chill, watched as the still bay faded from pink to indigo to black, and talked about our adventure and good fortune as we slipped through the night's deepening beauty for home.

———

People ask me what was the most important lesson I learned from the trainers and their animals. *To consider what you are reinforcing*, I answer. *To ignore what you don't like*, I add. Oh, and *to use incompatible behaviors* and *to know your species*. Also, maybe *not to act like a prey animal around a predator.* I can't make up my mind. Truth is, it's hard to say, to boil it down to the *one* thing. I suppose if I must, it is that the trainers and their animals showed me how to look at the world anew.

On a plane to Los Angeles, I notice that every time a baby cries, the father pulls it close and makes a kind of lilting shushing noise. The baby grows quiet. The father stops. The wee one pipes up again. "Shush, shush, shush," the father resumes. As I watch this behavioral cycle spin, I debate who's trained whom. Certainly the baby has got the dad shushing on cue to his crying.

I have Shamu moments like this all the time now. I see Al Gore on the big screen and worry that he is, despite his good intentions, desensitizing us to global warming, that like Kiara the lioness with her crate, we soon will be so accustomed to the idea of a coming climatic apocalypse that it won't scare us at all. I watch *Supernanny* on TV and notice her techniques are similar to the animal trainers', but she could stand to learn about approximations.

I understand better why people do what they do. When a friend hung in with a difficult boyfriend, I knew why. He had her on quite the variable reinforcement schedule. When my mother snarled at me when I suggested she get a hearing test, I knew I was the target of her displaced aggression, her rage against old age, and redirected her by cracking a joke and changing the subject. When any man clears his throat, I still find it disgusting, but I know they do it because rearranging phlegm, unfortunately, can be very self-reinforcing.

Last fall, as I staggered under the workload of writing this book, teaching, and preparing a speech, I remembered a scene from my visit to SeaWorld. I had stood for an hour watching a kind of pool party at the killer whales' tank. Trainers jumped onto big square orange floats, which the whales then pushed around with their great black foreheads. The trainers also dived into the water and grabbed hold of the whales' black dorsal fins. The orcas towed them around the deep blue tank. There was no apparent rhyme or reason to all this splashing, jumping, towing, and pushing. I learned later that this was a play session. The trainers thought the whales needed to blow off some steam, to just kick up their heels and have some fun.

I longed to push a raft around with my forehead. I desperately needed to blow off some steam. I had let my animal, me, get tired and frustrated. A trainer would never do that. Typically I would have soldiered on, but that image of the whales and trainers cutting loose together inspired me to do something different. I dived into the pool and took the afternoon off.

———

Obviously, thinking like an animal trainer has changed me, and in turn my marriage and life, for the better. But it has not made me rich, nor a genius, nor immune to bad luck. Fate still bats me with its huge paws, occasionally sinking its fangs into me. Life, like a wild animal, even a well-trained wild animal, remains fundamentally unpredictable. I have the scars to prove it, especially a fresh one, still tender to the touch.

When Dixie was five, we learned that one of her kidneys had shriveled. The other was failing. For one year we popped a small white antibiotic into her mouth twice a day. We fed her special food that would make her one working kidney's job easier. The vet said we might have her for months or years, it was hard to say. It always is with kidney disease. We hoped for the best. That was easy to do because our fireball dog continued to be a fireball, chasing down Frisbees, diving into cresting waves after tennis balls, ramming our legs with her head when she wanted a toss. Her amber eyes lit our life.

One morning, nearly three years after that diagnosis, she didn't lift her head when we walked into the room, not even when I picked up her leash for a walk. I rushed her to the vet, who sent us on to the emergency room, where I handed over my dying dog to a vet tech in scrubs. As she led Dixie down a tile hallway away from me, I watched my pup's withers sway like a grass skirt for what I thought was the last time.

Water was immediately piped into Dixie intravenously to flush out the toxins her kidney couldn't anymore. The clinic called the next day to say she was eating and alert. Three days later Dixie was ready to come home, but on borrowed time. Six months was the most we could hope for, the vet told us, and only if we could treat her at home. We drove home with an array of pills, a bag of sterile needles, and a box full of plump water pouches. Every day, we would have to pierce her skin and squeeze a half-liter of water under it.

We couldn't properly train Dixie for this procedure. There wasn't time for approximations, as we had to get a whole pouch of water into her that first day home. Still, training helped. Though we were nervous, we never let it show because if we did, then Dixie would know this bargain-basement dialysis was something to worry about. So we babbled and chirped. "Dixie dog, doodle bug," we'd sing. I'd kneel on the floor and ask her for a behavior she already knew, to push her nose against my balled fist. That way I could position her so that Scott could easily grab a handful of her ruff and stick her with the needle. I shoveled treats into Dixie's mouth throughout as she stood calmly, rarely even flinching, and a great water-filled welt rose on her shoulders. Afterward we tossed balls to her to reinforce that these nightly treatments were a good thing. Our plan worked. Every night we called, "Water," and every night Dixie obliged.

At first, Scott and I took turns jabbing the needle through Dixie's hide, which was no fun. Though I could do it, I was a little wild with the needle. I had to prick her quickly, without thinking about it too much, otherwise I might freeze. My method understandably made Scott anxious, especially after the time I stuck the needle through a handful of Dixie's hide and into my own palm on the other side. From then on, whenever I manned the needle, Scott tensed, which in turn would worry Dixie, and me too. He insisted on doing all the sticks. At first, I felt my inner primate swell. Who was he to tell me what to do! Then I realized that I needed to keep the nightly water sessions positive for Scott as well. I left the pricking to him.

In no time, Dixie was back to her maniacal self—pulling on her leash like a runaway horse when we headed out for a game of fetch and charging to the back door whenever we called "Car ride!" When I gardened, she'd drop a tennis ball repeatedly in the hole I was digging. In the park, people asked if she was a puppy. The vet took to calling her the Wonder Dog. Still, I didn't renew her license, and when I broke one of our two dog bowls, I didn't replace it. We took pictures of Dixie all the time, of the two dogs together, of the four of us. It was a summer and fall of last this and thats: her last swim in a kettle pond on Cape Cod, her last Halloween dressed as Batgirl, her last Christmas tearing open presents.

When the six-month mark came and went over the holidays, I began to wonder if Dixie would show us all, maybe she would see her ninth birthday, maybe even another Maine summer. I went to city hall and renewed her license. I let my mind drift to visions of tossing her a Frisbee across green grass.

The first day of February was gray, frozen, and forlorn. Dixie caught a ball that morning, even chewed a stick, though according to her blood tests she should have been dead days before. As the hours marched on, her gait grew unsteady. She stopped eating. She began to run into things. We scheduled the vet to come the next morning to put her down at the house. By evening, when the amber light faded from Dixie's eyes, it was clear we couldn't wait. The four of us loaded into the car and drove through an unreasonably cold night under the crisp light of a full moon. The traffic lights were green all the way. The only time we stopped was when a bus, the number 63 glowing in the dark, heaved to a stop to let its lone rider off. We waited as a hunched man in baggy jeans shuffled across the road in front of us. In the backseat, Dixie, her head hanging, struggled to sit up for her last car ride.

Scott, Penny, and I rose the next morning to a still house. Like a team that just lost their star player, we looked at each other and wondered how we would go on. Penny searched for Dixie in the house. Scott and I e-mailed friends, took flowers to the vet who'd seen us through, and cried endlessly. We staggered through that first day, and then the day after, and so on. Everywhere we went, everything we did reminded us of our golden girl. We longed for Dixie so, but just as much we missed the magic she had brought to our ordinary lives. The light that had shone so brightly on us was gone.

Thinking like a trainer could not turn the light back on, but it helped. I set about training myself to live without Dixie. I approximated my grief, pacing myself with first this and then that without Dixie. I slowly returned to the parks and beaches we had walked with her, starting with least favorites and working up. The big favorites, LeCount Hollow on Cape Cod, Sandy Beach on Long Island in Casco Bay, I saved for the summer.

I thought of incompatible behaviors to grieving so that I might get a break from my great sadness. I couldn't sob and shop at the same time, so I shopped with abandon. Likewise, Scott and I ate out whenever we wanted. The pleasure of the food and the bubbly crowd made it impossible, briefly, to mourn. After years of talking about buying tools and a stand so he could easily work on his bike, Scott did it. For him, cleaning gears and adjusting brakes was an incompatible behavior to missing Dixie.

For the first time, we turned our undivided attention to Penny Jane. For her five years, she had lived in Dixie's sizable shadow, happily I would say. When she was a scared, unsocialized pup fresh from the shelter, that shadow had, in fact, been her salvation. It wasn't that we hadn't trained Penny Jane. Our border collie mix had learned to be a very well-mannered dog who came when called and stood calmly while you hooked her leash. She also dug on command and had a mean high-five. But as the second dog, she hadn't been trained nearly as much.

Part of the great joy of having Dixie, I realized, was all the games we had trained her to play. Scott had taught her to play one-on-one soccer, during which he tried to dribble the ball with his feet as Dixie, madly circling him, tried to steal it. I taught her to find toys I hid in the house and to kick a ball to me with her front paws. That training was not only fun but it brought out Dixie's personality and intensified our bond with her. So I grabbed a ball and my clicker and began teaching Penny Jane to catch a Frisbee and fetch a tennis ball. Scott, kneeling on the floor, trained Penny Jane to put her front paws on his shoulders and take a treat from his mouth, which always cracked us up.

We went to the park and taught Penny Jane a game we had long ago first played with Dixie. We walked ten feet apart and sent her back and forth between us. "Go to Amy," Scott told her, and Penny Jane, her mouth wide in a smile, tongue flapping, front paws flying, ran to me. "Go to Big Dog," I'd say, and she'd turn, dig in her white paws, and sprint back to Scott. With each turn, Scott and I took a few steps backward, until we were so far apart that we couldn't hear each other's commands. I saw Scott point to me, and Penny again hurled herself in my direction. When I sent her back to Scott, as I watched the flash of her curled tail recede from me, a small white curlicue against the blue sky, I smiled, and in that moment, my grief lifted and the magic of Penny Jane, of two people mending their broken hearts by teaching a dog to run back and forth between them, of the great joy of our deepening threesomeness, took its place.

As I said at the start, the world is full of surprises. And maybe, in the end, that's the most important lesson I learned from the animals and their trainers.

ACKNOWLEDGMENTS

This book owes the animal kingdom a huge debt. That includes a long list of human animals, starting with the one I'm married to, Scott Sutherland. It is a brave man who marries a journalist, a braver one yet who lets his author-wife write about him and his quirks. My husband is as ballsy as a big-cat trainer, not to mention that he'd look fine in one of those tight-fitting sparkly suits. The best way of thanking him would be to promise to never write about him again, but being a writer I can't. All I can promise is to love him the rest of my days. I promise.

Next up, my editor at Random House, Stephanie Higgs, deserves a bucket of mackerel—actually, a truckload of fresh tunas. Among other good deeds, she took the wild animal of my first draft and trained it to walk nicely on a leash. Buckets of fish also to all staff members at Random House who made this book happen.

Mackerels galore to my literary agent, Jane Chelius; my film agents, Mary Alice Kier and Anna Cottle; and my publicist, Megan Underwood Beattie. I am a lucky author in that I can say that I'm good pals with all of them. Everyone should have such smart, funny women on their side.

As this book is intricately linked to my previous book, I find myself needing to thank some of the exact same people I did then. Dr. James Peddie opened a gateway to a world that has not only provided fodder for two books, but has led me on a journey that has truly transformed me for the better. I also owe another round of thanks to the talented staff at Moorpark College's Exotic Animal and Management Program, especially Mara Rodriguez, who answered many an inane question via e-mail for this project.

Likewise, I once again relied on the knowledge of professional trainers. I drew heavily on the writings of Steve Martin of Natural Encounters, Inc.; *Animal Training: Successful Animal Management Through Positive Reinforcement,* by Ken Ramirez of the Shedd Aquarium; and *Don't Shoot the Dog!,* by Karen Pryor. In fact, this book wouldn't be possible were it not for Pryor, who was a leading pioneer in the revolution that has transformed animal training in this country.

As always, I owe my family and friends big-time. Both my mother, Joan, and brother, Andy, were good enough sports to let me turn the animal training lens on them. My author pal, the brilliant Hannah Holmes, was the first to see this idea as a book, before it even became a column. Friends Dana Baldwin, Nancy Bless, Becky Stayner, and Elise Williams not only let me include them in the book, but saw me through my darker moment, when all this dolphin wanted to do was sink to the bottom of the tank.

Less than two years ago, I hit a key on my computer and sent a column about using exotic animal training to improve my marriage over the ethers to *The New York Times*. Before long, a reply from editor Daniel Jones landed in my in-box. His expert editing gave birth to the column that became this book. Mackerels to Daniel and *The New York Times*.

That brings me to the animal animals. So many have shaped my life: my first dog, the stick-loving Curly; Schmoo, the dowager sea lion at Moorpark; the pair of blue jays that have recently deigned to dine at my urban feeder; and, of course, my Maine mutt, Penny Jane. But none knocked my life for a loop the way my gorgeous Australian shepherd, Dixie Lou, did. At an age when I was well past believing in or even caring about magic, it bounded into my life in the form of a forty-five-pound dog with amber eyes. With Dixie, I found a new way to look at the world, a new way to relate to all living creatures and a new, better me. Moreover, she filled my everyday life, even the dullest moments, with such joy. Now that she's gone, I don't know if I'll ever be happy like that again, but, at least, I was.

GLOSSARY

A TO B: Training an animal to go from point A to point B on command.

AFFECTION TRAINING: A term used in Hollywood to describe training with positive reinforcement. Frank Inn, of *Benji* fame, was a practitioner.

BEHAVIORAL ENRICHMENT (B.E.): Anything that enriches a captive animal's life, from toys to walking on a leash.

CLICKER TRAINING: Essentially, dolphin training for dogs. Trainers use positive reinforcement plus a marker to tell dogs when they have given the right response. Instead of sounding a whistle, as dolphin trainers do, trainers click a metal noisemaker called a cricket or a clicker—thus the name *clicker training*.

COUNTERCONDITIONING: Making something bad into something good by pairing it with positive reinforcement. A form of desensitization.

CRITERIA: The specific training goal—for example, how high exactly you want the dolphin to jump.

DESENSITIZATION: Accustoming animals to new and potentially unnerving experiences, things, and places.

DISPLACED AGGRESSION: Venting rage on an innocent bystander or object.

EXTINCTION: When a behavior is unreinforced for so long it goes bye-bye.

GENTLING: The circus-world term for training with positive reinforcement. Big-cat trainers Mabel Stark and Louis Roth (her mentor and then her husband) were both reportedly proponents.

GOING BACK TO KINDERGARTEN: Taking a few steps back in the training process.

HABITUATION: Desensitizing an animal to something potentially unnerving solely through exposure.

INCOMPATIBLE BEHAVIORS: Behaviors that make unwanted behaviors impossible. A killer whale can't present a flipper poolside and pester another killer whale at the same time.

INNOVATIVE TRAINING: Teaching an animal to create new behaviors itself on cue.

INSTINCT: Innate or natural behavior, which typically is beneficial to a species. For example, pigs instinctively root, raccoons instinctively wash their food, and chickens instinctively peck at objects.

JACKPOT: An extra big serving of positive reinforcement—a bucket of fish, a handful of kibble, an entire sweet potato—for when an animal goes above and beyond.

LEAST REINFORCING SCENARIO (LRS): When a trainer, faced with an animal offering an incorrect response to a cue, does not respond at all for a few beats so as not to encourage that behavior. A neutral way of telling an animal, "Wrong."

LURING/BAITING: Showing an animal its reward in advance of a behavior—for example, holding a carrot out to lure a rhino to walk through an opened gate.

MARKERS/BRIDGES: Signals that tell an animal exactly when it gets a behavior right. Dolphin trainers use whistles. Dog trainers use clickers. Some trainers use a specific word.

NEW-TANK SYNDROME: Doing a learned behavior in a new setting can make that behavior deteriorate.

OPERANT CONDITIONING: A school of psychology based on the concept that learning is shaped by its consequences.

PATTERNED: When something is done simply out of habit.

PLASTIC BEHAVIOR OR PLASTICITY: Flexible or adaptive behavior, which makes animals quick learners. Social animals, such as dolphins and primates, typically have very plastic behavior.

POSITIVE REINFORCEMENT: Something an animal desires, which thus can be used to increase frequency of a behavior.

PRECURSORS: The events or circumstances that precede or prompt a behavior.

PUNISHMENT: Something an animal would rather not have anything to do with. Punishment is used to decrease the frequency of a behavior.

REHEARSED AGGRESSION: The more an animal aggresses, the more likely it will aggress again. Practice makes the animal better at aggression, which makes it more inclined to use aggression, especially if it wins the fight and gets what it wants, such as not having to go into its enclosure or scaring the trainer away.

SCENARIO: All the precursors and antecedents that prompt and/or surround a behavior.

SELF-REINFORCING BEHAVIOR: Behavior that in itself is enjoyable and thus reinforcing.

SHAMU: As a noun, the name SeaWorld coined for its killer whales. SeaWorld's very first Shamu was caught in Puget Sound in 1965. As a verb, see page 27.

SUCCESSIVE APPROXIMATION: Teaching by using incremental, sequential steps.

SUPERSTITIOUS BEHAVIOR: Behavior that results from accidental reinforcement, either by chance or because of a trainer's mistake.

TARGETING: Teaching an animal to press part of its body, typically its snout, against an object.

THROWING BEHAVIORS: When an animal offers behaviors willy-nilly to see if anything will earn it a treat.

VARIABLE REINFORCEMENT SCHEDULE: Reinforcing a behavior occasionally and on an unpredictable schedule. Used to maintain learned behaviors.

PHOTO: JAY YORK

AMY KILLINGER SUTHERLAND is the author of *Kicked, Bitten, and Scratched* and *Cookoff*. Her articles have appeared in *The New York Times, Los Angles Times,* and *The Boston Globe*. She has a master's degree in journalism from Northwestern University. Her feature piece "What Shamu Taught Me About a Happy Marriage," on which this book is based, was the most viewed and most e-mailed article of *The New York Times* online in 2006. Sutherland lives with her husband, Scott, and their dog, Penny, in Boston.

ABOUT THE TYPE

This book was set in Walbaum, a typeface designed in 1810 by German punch cutter J. E. Walbaum. Walbaum's type is more French than German in appearance. Like Bodoni, it is a classical typeface, yet its openness and slight irregularities give it a human, romantic quality.